A History of the Wind

Alain Corbin

Translated by William A. Peniston

polity

Originally published in French as *La Rafale et le zéphyr: Histoire des manières d'éprouver et de rêver le vent* by Alain Corbin © Librairie Arthème Fayard, 2021

This English edition © Polity Press, 2023

Polity Press
65 Bridge Street
Cambridge CB2 1UR, UK

Polity Press
111 River Street
Hoboken, NJ 07030, USA

ISBN-13: 978-1-5095-5205-4

A catalogue record for this book is available from the British Library.

Library of Congress Control Number: 2022940495

Typeset in 11 on 14 pt Sabon LT Pro
by Cheshire Typesetting Ltd, Cuddington, Cheshire
Printed and bound in Great Britain by CPI Group (UK) Ltd, Croydon

The publisher has used its best endeavours to ensure that the URLs for external websites referred to in this book are correct and active at the time of going to press. However, the publisher has no responsibility for the websites and can make no guarantee that a site will remain live or that the content is or will remain appropriate.

Every effort has been made to trace all copyright holders, but if any have been overlooked the publisher will be pleased to include any necessary credits in any subsequent reprint or edition.

For further information on Polity, visit our website:
politybooks.com

A History of the Wind

Contents

v

Contents

Translator's Note

Fundamentally, this book is a history of the imagery of the wind in literature, from the Bible and the ancient Greek and Roman epics, through the Renaissance and the Enlightenment, to the modern era. There is a wonderful section on the mythology of the early modern nations and a heavy emphasis on nineteenth-century Romanticism. For the most part, Corbin uses modern editions for the French writers and modern translations for the sources originally published in Portuguese, Italian, German, and English. Whenever possible, I have used existing translations of the foreign-language materials, and, of course, I have tracked down all the original quotations for the English and American citations. Given my limited access to libraries during this time of social isolation, I have relied on online sources, especially the Internet Archive and the Hathri Trust. Consequently, for better or worse, I have cited eighteenth- and nineteenth-century publications.

I would like to thank the anonymous reviewers who help polish the final draft of this translation.

George Robb read very carefully the initial draft and gave me some very thoughtful suggestions for improvements.

William A. Peniston, September 2021

Acknowledgments

I thank Sophie Hogg-Grandjean and Pauline Labey for having assisted with the development of this manuscript, and Sylvie Le Dantec for having accepted it.

Alain Corbin, April 2021

Prelude

In the nineteenth century, scientists began to understand the wind. Before then, this noisy emptiness was experienced and described only according to the sensations that it provoked. Inconsistency, instability, and imperceptibility defined this invisible, constant, and unseen flow. The fleetingness of the wind – and the immense scale of its power – explained why not much was known about where it came from and where it went.

Everyone could experience its presence, its force, and its influence: the wind blew at times, and sometimes it cried, roared, or howled. It was, above all else, the sound and the fury. Occasionally, it seemed to moan and groan like a soul in pain condemned to an eternal damnation. Its energy aroused dread: it assailed, it brutalized, it whipped up, it knocked down, and it uprooted. That is why it was identified with anger. Furthermore, it swept away, it carried off, and it dispersed things in its flight. It both dried the countryside and fanned the flames of fire. Nevertheless, the wind also sighed, caressed, and seemed sometimes to play the role of the lover.

The wind's action on man's body is contradictory: here it freezes, there it stifles. Since antiquity, it was thought that it purified and improved health, but it could also stink and poison in the literal sense. Briefly, the wind – what Victor Hugo called "the sob of vast expanses, this breath of space, this respiration of the abyss"[1] – could, in the course of time, arouse fear, dread, and hatred.

These notions give rise to the thought that the wind with its immutable traits has escaped history. This is far from being the case. Beginning to understand it, however, being persuaded of its far-off origin, and perceiving its mechanisms and its trajectories are all historical facts dating to the dawn of the nineteenth century. The same is true of new experiences of the wind at the summit of a mountain, or in the desert, or in the depths of an immense forest, or, more than any place else, in the air.

Moreover, at the same time, the ways of perceiving and feeling the wind were enriched by the rise of a "meteorological self." From that time onward, the wind as a literary object has not ceased to inspire writers. The ways of imagining it, or speaking about it, or dreaming about it were modified and enriched by the code of the sublime, by the exaltation of nature in German poetry, and by Romanticism itself. And let us not forget the successive reinterpretations of the wind in the epics that have, in the course of the centuries, conferred upon it an essential place in modern culture.

It is necessary to highlight right from the start the level of knowledge or ignorance concerning the wind if we are to understand clearly the various ways of experiencing it. That is why we will begin by reviewing the scientific turn that took place at the very end of the eighteenth century, notably the discovery of the composition of air. We will then describe the new understanding of atmospheric circulation and the new experiences involving the wind. We will not neglect the aesthetic forms that governed the emotions aroused by this elemental force.

After having placed the wind at the heart of these experiences, we will explain in rough outline the manner in which artists, writers, and travelers have, since antiquity, interpreted and, above all, dreamed about this force that has no equal, this indecipherable enigma that was formed by the wind. These references are related to the new knowledge and the new experiences that brought about a renewal of the imagination in the eighteenth and nineteenth centuries.

In summary, an immense field of research is sketched out for the historian, all the more so since the wind is also, perhaps above all else, a symbol of time and oblivion. That is why we should meditate on Joseph Joubert's formula: "Our life is made of woven wind."[2]

1

The Inscrutable Wind

During the night of July 4–5, 1788, Horace Bénédict de Saussure (1740–1799), who had climbed Mont Blanc the year before, experienced a wind with an intensity until then unknown when he made an excursion to the Col du Géant [Giant's Pass]. It seemed to him to be so new that he described it in detail in his book *Voyages dans les Alpes*. Having taken refuge in a little cabin with his comrades, he wrote:

> This wind from the southwest rose up at an hour after midnight with such violence that I believed it was going to carry away at any moment the stone cabin in which my son and I were sleeping. It was one of such uniqueness that it was periodically interrupted by intervals of the most perfect calm. In these intervals, we heard the wind blow below us in the depths of the Allée Blanche, whereas the most absolute tranquillity reigned around our cabin. But these periods of calm were followed by gusts of an inexpressible violence; they were repetitive blows, like artillery fire, or so it seemed. We felt as if the mountain itself was shaking

under our mattresses. The wind came in through the cracks in the cabin's stones; it even lifted up my covers and froze me from head to toe. It calmed down just a little at the break of dawn, but it picked up again soon and returned with snow that entered every part of our cabin. We then took refuge in one of the tents ... There we found that the guides were obliged to hold up the poles continually out of fear that the wind would knock them down and sweep them away, along with the tent.[1]

Saussure then described the "hail" and the "thunder" that assailed his party.

In order to give an idea of the intensity of this wind, I will say that twice our guides, wanting to check on the men in the other tent, chose an interval when the wind seemed to have calmed down. Half way there, despite the fact that there were only sixteen or seventeen steps between the tents, they were struck by a gust of wind so strong that, in order not to be carried over the cliff, they were obliged to cling to a rock that was happily located nearby. They remained there two or three minutes with their clothing flapping over their heads by the wind and their bodies riddled with hailstones until they could resume their task.[2]

In that summer of 1788, such an experience of the wind seemed new to Saussure, even though it must appear very ordinary to readers today. It is this assessment that is a historical fact, and we will see that, in the decades to come, there were other experiences of the wind that seemed new. At the end of the eighteenth century, when the study of air was becoming fashionable, the wind was still seen, for the most part, as an element. Several decades later, the nature, the origin, and the circulation of the wind would be better understood.

Until the end of the eighteenth century, we possessed very few scientific facts about the wind. New experiments undertaken in the course of navigation or, periodically, during the exploration of different regions added new proofs, often terrible proofs. For the most part, the wind was described in its local manifestations, and we will come back to this point. Sailors accorded it an extreme importance; they used a number of words and expressions to describe it. Its nature was understood, however poorly, through the methods of recording it used by amateurs who had at their disposal various measuring instruments: the anemometer, the thermometer, and the barometer sometimes figure in the small laboratories of these men who were passionate about meteorology. And we should not forget about the installations of weathervanes, designed to indicate the wind's direction and placed on the towers of churches and castles, because they were a sign of feudal privilege.

The most educated individuals continued to perceive the wind according to the religious and literary representations that emerged from the mists of time; that is to say, the wind was perceived as an essential fact of human life, and yet it remained inexplicable. Certainly, since the Renaissance, navigators had discovered the regular circulation of the trade winds in the tropics, and some sea charts from that era took into account these observations. Moreover, certain local winds, such as the mistral, the tramontana, and the northwester (maintaining our focus on France), had been described with great precision. And let us not forget that, at the end of the eighteenth century, in the salons, small demonstrations were held in the course of which scientists, or so-called scientists, reproduced the blowing of the wind. However, to understand this last fact requires knowledge of air and its composition. Was it an elemental fluid, alongside water, earth, and fire, as was thought since Aristotle's time? Or was it a mysterious phlogiston?

Whatever it was, from then on, the specialists thought that air acted in many different ways on the body: through simple contact with the skin or with the pulmonary membrane, through exchange across the pores, or through direct or indirect ingestion since food contained it. Scientists of this time stated repeatedly that, according to the seasons or the regions, air regulated the tautness of the body's fibers; this was an essential fact. It was observed that a precarious balance was established between external air and internal air in the body, constantly restored by exhalation, expectoration, belching, or the "winds." Two centuries earlier, Rabelais had invented the Island of Ruach, whose inhabitants lived off nothing but the winds.[3]

All of this contributed to the conviction that air was animated by a *force*, an elasticity, large enough to equal the force of gravity. In this perspective, when air loses its elasticity, only the movement of it, its agitation – and we will return to this point, too – could restore it and thus permit the survival of the organs. In the opinion of doctors, this balance between the body, the internal organs, and the atmosphere constituted an essential fact: hot air caused an elongation and a relaxation of the body's fibers, whereas cold air caused a tightening of the fibers, and fresh air was revealed to be particularly beneficial. It should therefore be studied. And so it is understandable how these scientific representations of air constituted the foundation for an interest in the winds.

This concept of air led to the idea of air as a frightening broth in which smoke, sulfur, and vapor, whether watery, oily, or salty, were all mixed together. Indeed, all inflammable materials exhaled by the earth, such as the emanations from swamps and the miasma arising from decomposing bodies, were suspect. All of this compromised the wind's elasticity, sometimes through the strange fermentations and transmutations that were accompanied by thunder, lightning, and storms.

The atmosphere of a place constituted a menacing cistern from which epidemics risked breaking out. All of this led to the conviction that the wind, the agitation of the air, was capable of clearing out its noxious charges. Neo-Hippocratism – the doctrine dating to Hippocrates in the fifth and fourth centuries before the Common Era and laid out anew in the eighteenth century – led to the advocacy of an atmospheric vigilance and a mistrust of times of great calm. The praise of ventilation extended across a long period of time: from the period during which air was conceived as an element or a phlogiston to the era when it was believed to be composed of chemicals. And now it is this concept that we must consider.

The concept of phlogiston was once considered one of the greatest forces in nature. It was thought that it comprised a particular fluid, inherent in all beings, one that produced combustion when it abandoned the body. This theory, sketched out in the seventeenth century, had been taken up again and developed by Georg Stahl (1660–1734), one of the most eminent scientists of his time. According to him, phlogiston existed in all combustible bodies, and combustion itself was only the passage of phlogiston from a combined form into a free form.

Antoine Lavoisier (1743–1794), as we know, destroyed this false interpretation of combustion. Like Joseph Priestley (1733–1804) – and we are going to come to him later – who had described the composition of air in his own way while still remaining caught up in his fidelity to the idea of phlogiston, Lavoisier demonstrated that air was composed of nitrogen – as acknowledged by Daniel Rutherford (1749–1819) since 1772 – and oxygen and hydrogen – as identified by Henry Cavendish (1731–1810) in 1781.

Priestley's discoveries, published in 1772 and 1778, were important but incomplete. According to this scientist and clergyman, in studying respiration, there was a "common air," a "phlogisticated air" (nitrogen), and a "dephlogisticated air" (oxygen), which was "vital." This last kind of air was excellent for breathing. In brief, then, Priestley's partial fidelity to the idea of phlogiston prevented him from succeeding in describing perfectly the composition of air. Nevertheless, in his work, air ceased being an element and began being perceived as a combination or a mixture of gases. According to Priestley, along with other scientists from his era, the chemistry of gases and organic processes were directly linked together. To study the different kinds of air was to study the mechanisms of life, and to ventilate public spaces was to purify them. It was then understood that wind was at the center of this emerging concept of public hygiene. Ventilation was thus the axis of the hygienic strategy, since it was guided by the fear of stagnation and fixity.

Even before Lavoisier's discovery of the exact chemical composition of air, the neo-Hippocratic understanding of air led to the advocacy of ventilation as a restorer of the elasticity of air as much as an antiseptic. The wind swept away the hidden parts of the atmosphere and purified and deodorized the polluted waters. In a word, to survey and to master the winds and the currents was considered an essential practice.

With this perspective in mind, the bellows and all ventilators revealed their usefulness. Many objects were capable of stimulating the beneficial effects of wind – that is, the circulation of air – or so it was thought: fans in private places, trees next to marshes, rotating windmills on sleds, vehicles of all kinds inside cities, the disturbance of the atmosphere by bells, the explosions of cannons, sails on ships, and so on. In the hulls of ships, merchandise suspected of transmitting diseases was ventilated.

The architecture of the Enlightenment was obsessed with the need to circulate air and to create rising currents of air. A healthy town should not be surrounded by walls because they would hamper this means of purification. Streets should be wide and squares vast in order to encourage the circulation of the winds. Similarly, it was advisable for buildings to be distant from one another, and hospitals were conceived as "islands in the air." For example, the goal of a decree by Louis XVI was to promote the ventilation of spaces and the circulation of air inside cities.[4]

In England, as in France, royal societies of medicine extolled the creation of the "medical regime" in several localities. In particular they advocated the better detection of sanitary conditions, especially in regard to the risks of epidemics. This practice, in the French case, was presented in the famous report of departmental statistics by Jean-Antoine Chaptal (1756–1832) under the Consulate and the Empire, and it continued to be the object of innumerable brochures during the first third of the nineteenth century. It constitutes an important element in the history of local winds. All of this is very well known.

Let us return to the key theme in question. Between 1800 and 1830, knowledge about the wind progressed only very slowly. Nevertheless, we can note two major concepts which emerged during this period. Scientists became convinced of the existence of an "aerial ocean," while awareness of the distant geographic origins of its related phenomena grew. These two concepts bore the mark of one of the greatest scientists of the era, Alexander von Humboldt (1769–1859). Let us cite a couple of passages from his great book *Cosmos*, which was published in 1845 and summarized his earlier thought. Having outlined the contemporary knowledge concerning the "aerial ocean," he wrote:

The second and external and general covering of our planet, *the aerial ocean, in the lower strata, and on the shoals of which we live*, present six classes of natural phenomena, which manifest the most intimate connection with one another. They are dependent on the chemical composition of the atmosphere, the variations in its transparency, polarization, and color, its density or pressure, its temperature and humidity, and its electricity.[5]

A few pages further on, Humboldt highlighted this fact, which was essential in understanding our subject, the wind. He stressed geographic distances as causes for atmosphere events. From him we learn:

> Important changes of weather are not owing to merely local causes, situated at the place of observation, but are the consequence of a disturbance in the equilibrium of the aerial currents at a great distance from the surface of the earth, in the higher strata of the atmosphere, which bring cold or warm, dry or moist air, rendering the sky cloudy or serene, and converting the accumulated masses of clouds into light feathery *cirri*. As therefore the inaccessibility of the phenomena is added to the manifold nature and complication of the disturbance, it has always appeared to me that meteorology must first seek its foundation and progress in the torrid zone, the course of hydro-meteors, and the phenomena of electric explosion, which are all periodical occurrences.[6]

As we can well understand, these cautionary lines were written on the cusp of the discoveries which we will shortly examine.

Nevertheless, Humboldt was not the only one who emphasized distance in the creation of the winds, rains, and atmospheric currents. In his own way, in a much more poetic fashion, Jacques-Henri Bernardin de Saint-Pierre (1737–1814) demonstrated a fascination in their origin and dreamed of their future progress.

The understanding of meteorological phenomena, such as the wind, began to develop abruptly after 1854 and 1855. In those two years, two natural disasters left a distraught public in their wake. On November 14, 1854, a terrible storm struck the English and French fleets near the Crimean peninsula; a number of vessels were destroyed, including the *Henry IV*, the jewel of the French navy. And on February 16, 1855, the frigate *Sémillante* sank without any survivors. Emperor Napoleon III was very shocked and made several decisions regarding naval safety. That same year, Urbain Le Verrier (1811–1877) became the director of the Observatory in Paris, where an employee began to transcribe the direction of the wind in a register three times a day. In Greenwich, as in Paris, scientific publications multiplied, and the networks of observations grew more and more dense. International conferences were organized, and one of them, a gathering of experts from ten different countries, called for meteorological observations to be taken several times a day on board ships. At the same time, the public became interested in the dynamic meteorology of the air. Soon, in 1859, the underwater telegraph accelerated the transmission of facts.

All of this research stimulated scientific discoveries, and those concerning the wind followed one after another. Already in 1848, Henry Piddington (1797–1858) of Calcutta published a reference work on tropical storms entitled *The Sailor's Horn-Book for the Law of Storms*. He introduced the term "cyclone" (from the Greek for "circle") in order to describe those circular storms. The word was adopted into French eleven years later. In the same year, 1848, Matthew Fontaine Maury (1806–1873) of the American navy produced his study, *Wind and Current Chart of the North Atlantic* (figure 1). In 1863 Francis Galton (1822–1911) introduced the concept of the "anticyclone," while the Dutchman Christoph Buys Ballot (1817–1890) put forth the law that explained the direction of the wind in relation to the center of the cyclone.[7]

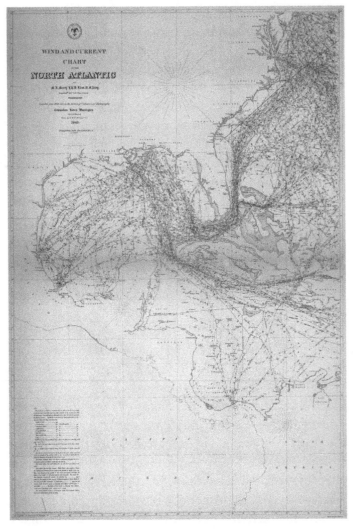

Figure 1 Matthew Fontaine Maury, *Wind and Current Chart of the North Atlantic*, 1848.

© American Geographic Society Library – Maps. From the American Geographical Society Library, University of Wisconsin-Milwaukee Libraries.

Dynamic meteorology continued to develop in France as well. Édme Hippolyte Marié-Davy (1820–1893) played an essential role in this field. First of all, he detected what he called "cyclonoids," storms of a vast diameter, which he soon renamed "squalls." In 1863 he published several maps. Two years later, having abandoned the idea that these were only tropical storms that had grown weaker as they pursued their course toward Europe, he pointed out that these "squalls" originated in the regions of the New World, in Iceland, and in the Azores. They took several days to reach Europe. In the course of the 1870s, these "squalls" became known as fundamental entities of atmospheric dynamics, especially in their effect on Europe.

At the same time, there sprang up a veritable passion for wind maps. Between 1848 and 1873, Maury and his staff thus developed and published "wind rose" charts, in which one could read the number of times the winds blew and the directions they took for each month of the year, all based on direct observations.

The most enthusiastic scientist studying the directions and the intensity of the winds was, without a doubt, Léon Brault (d. 1885), a lower-middle-class Catholic librarian. For a project assembled in 1870, he visited various ports in order to discover in their archives numerous facts concerning the winds. His goal was to detect "the normal equilibrium of the atmosphere" and not just the accidental ones made up of tropical storms, cyclones, squalls, or depressions, which he considered "sicknesses of the atmosphere." In 1873, following up on his endless travels, and thanks to the work of an extensive team, he succeeded in producing twelve notebooks gathering together, in total, 750 observations as recorded by sailors from different ports. He ranked these observations according to a traditional scale developed by the French navy, and not according to the progressive, numbered scale of Francis Beaufort (1774–1857), which was adopted by the British Royal Navy in 1838. The anemometer that he used was the bodies of the sailors themselves, not a scientific instrument. In effect, the facts gathered together had been collected by officers of the wheelhouse whom Brault considered to be the best anemometers of all.[8]

At first, at an international conference in August 1875, Brault presented his statistical maps of the winds of the North Atlantic (figure 2). According to him, his work settled the disputes over the winds dating from the end of the eighteenth and the beginning of the nineteenth century. He deemed these disputes to be founded on an *a priori* knowledge. In 1877, and again in 1880, he published quarterly maps of the winds of the South Atlantic and then of the Pacific and Indian oceans. He then began a similar project on the ocean currents, but this was interrupted by his death in 1885. For decades, Brault's maps enjoyed a tremendous success in scientific circles. Until 1940, they were used in the construction of the entire fleet of the French navy.

Let us return to the ocean currents, because we know today that wind plays an essential role in their existence. However, what was known of them then? The Gulf Stream – its existence and its itinerary – had been known for a very long time. Humboldt dedicated a very precise study to it. He mentioned the role played by the "tides [that] occur in their progress round the earth; the duration and intensity of prevailing winds; the modifications of density and specific gravity which the

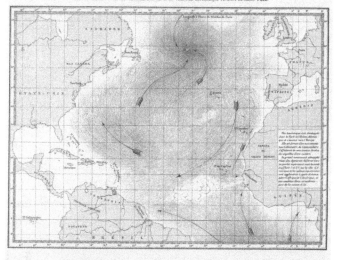

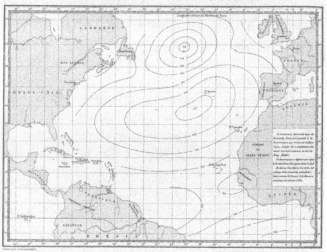

Figure 2 Léon Brault, *General Movement of the Winds during the Summer Season*, 1885.

Department of Cartography, Navy of the Republic of France.
© Bibliothèque Nationale de France.

particular waters undergo in consequence of differences in the temperature and in the relative quantity of saline contents at different latitudes and depths."[9] He also emphasized the variations in atmospheric pressures. And, finally, he described the movement of these currents and their velocity and posed the problem of their depth.

In 1855 Maury, in his work *The Physical Geography of the Sea*, published maps of the winds and the currents, contending that Brault's maps varied a great deal from one another. It was at this time that the discovery of deep currents took place, but it was not until much later that the role of the wind was precisely described in this area. In fact, we now know today that close to 50 percent of the energy exercised by the wind on the ocean is transferred to the currents. As Jean-François Minster recalled in his *La Machine océan*, "the geographic structure of the wind is responsible for the horizontal structure of the currents on the surface and for their vertical displacement."[10]

The interest shown in the winds by scientists – and by the public at large – found its crowning achievement in the aggregation of the science of the wind in the geography of Paul Vidal de La Blach (1845–1918). At the end of the nineteenth century, thanks to the knowledge acquired about the dynamics of air, the winds ceased to be unknown. However, knowledge of the upper atmosphere was still lacking, notably the concepts of the troposphere and the stratosphere, which were introduced by Léon Teisserenc de Bort (1855–1913) at the beginning of the twentieth century. In addition, knowledge about the jet stream became widely known in the middle of the last century. In

France, the meteorologist Pierre Pédelaborde played a major role in the field at the end of the 1950s, notably through his teaching at the University of Caen, where I was one of his students.

2

The Winds of the Common Folk

Knowledge of the laws governing the distant atmospheric circulation of air held little relevance to most people's everyday experience of the wind. It was the local winds, the learning of their names and the notions concerning them, transmitted from generation to generation, which were important. These have been studied time and time again. Very precise lists of information had been established for the winds of each region. It would be tedious to get lost in all of this detail.

As an example, though, Jules Michelet (1798–1874) inventoried the four winds of the south – the foehn, the autan, the sirocco, and the simoom – which assailed the mountain that he described. In France, the most famous of all local winds is the mistral, which can be cold, dry, and violent. It blows from the north to the northwest in gusts that reach 60 miles an hour. It is found in the triangle formed by Valence, Montpellier, and Fréjus and is most intense in the corridor of the Rhône Valley.[1] It has a major impact on nature: it encourages vegetation; it erodes cliffs; it accentuates the light of the sky; it dries and it causes fires; and it rips off the roofs of houses. In general, it blows for one to three days and, in spreading over the sea, can reach Corsica and the Balearic Islands. No doubt it caused the shipwreck of the *Sémillante* in February 1855 off Bonifacio. The rural architecture of these regions takes it into account. Evidently, until the middle of the nineteenth century, it enjoyed a great popularity among windmill owners.

To paraphrase the words of Jean-Pierre Destand, the author of an article on the "aeolian winds of Languedoc," the local winds mark the boundaries of territories and native areas. Their diversity is reflected in a range of names, each of which has its own emblematic value. They play the role of living domestic barometers. Hence, they exercise a role in the forms of spatial organization.[2]

Often the wind is a reference point. It governs strategies of fishing, harvesting, and hunting. For this reason, we can't help but pay attention to it. Destand distinguishes ordinary winds, simple crosscurrents of air, recurrent breezes, "little winds," and "light winds." He lists the forms of "aeolian knowledge" and "aeolian culture" of the local inhabitants. He states that the wind is often the subject of their conversation. They wish for it, they implore it, or they curse it. It is a regrettable presence or a regrettable absence. Sometimes, when it stops, it instills a feeling of loss. For some, it fills a void with its blowing; and, for others, it makes them feel the silence that is imposed by its interruption.

Each wind has its own tactile manner of making itself felt – and we will come back to this further on. There are some winds that sing and others that whistle. Certain ones, more than others, spread scents or odors.

All participate in the creative poly-sensorial nature of the countryside. Some winds cause accidents; motorists are well aware of this. There are winds of places as much as winds of peoples, according to Destand. One of his sources assured him that the wind on television was not one at all, because, according to him, true wind is the business of light, sounds, seaborne odors, or earthly scents. Their directions are defined and determined by visual markers.

Local winds have not ceased to capture the attention of historians. Martine Tabeaud, Constance Bourtoire, and Nicolas Schoenenwald undertook an analysis of their influence on, or rather their transposition into, children's literature.[3] Indeed, from very early on, the winds have weighed heavily upon human imagination. Patrick Boman, for example, specializes in establishing a list of "rain winds" in France.[4] The historians of weathervanes, such as Jean-Pierre Richard, have worked along the same lines, and so have those who have written the histories of windmills.[5]

On this point, let us consider what Alphonse Daudet (1840–1897) had one of his characters say in the story entitled "Master Cornille's Secret." Evoking the past, when farmers living 10 leagues away carried their grains to the mills to be ground up, he wrote: "The hills all about the village were covered with windmills. To right and left, one saw nothing but sails twirling to the mistral above the pines, strings of little donkeys laden with sacks going up and down the hills and roads." It was a pleasure to hear from the hilltops "the cracking of the whips" and "the creaking of the canvas." "On Sundays we went to the mills in parties. The millers, they paid for the muscat. Their wives were as fine as queens …

Those mills, you see, they made the joy and the wealth of our parts."[6]

In a letter to Hippolyte de Villemessant, which formed part of this book, Daudet related the story of a sleepless night in an old broken-down windmill. It was "the mistral [that] was angry and the roar of its great voice kept me awake till morning. The mill creaked, heavily swaying its mutilated wings, which whistled to the north wind like the shrouds of a ship. Tiles flew from its roof." Full of seawater, the large gusts of wind fell heavily against the door and made the hinges cry. Then Daudet thought of the sailors assailed by this gale that roared in their masts and tore their sails to tatters. "I might have thought myself on the open sea."[7]

In highlighting the extreme importance of local winds, the object of this chapter, it would be tedious to draw up a list of references. We will content ourselves with a review of this or that wind out of numerous others in the chapters that follow, and we will, therefore, have the opportunity to understand better the manner in which they have been experienced and the emotions that they arouse.

3

The Aeolian Harp

During the second half of the eighteenth century, there was a sudden surge of literary self-expression: intimate journals, notebooks, and correspondence, all attesting to an ever-intensifying "I" – an "I" that was sensitive to and reflected the weather. The risks to this "I" mirrored the risks of fate. And the history of the wind transcribed this new sensibility. This concept of the wind expressed and symbolized the sounds of nature, which have haunted literature ever since. "The tempest, the thunderstorm, and the hurricane all entered literature as signs of the basic instability of the human condition," especially in the guise of solitary subjectivity.[1] The *Sturm und Drang* movement illustrated this point in its formula of the "storm" as portrayed as an extreme intensification of the wind, most often accompanied by snow and associated with notions of assault, impulse, and momentum characterized by the word "stress."

A musical instrument quickly became the sign of phenomena larger than itself; the Aeolian harp symbolized this new attention (figure 3). Its first incarnation

Figure 3 An aeolian harp.
© Granger/Bridgeman Images.

was as a wooden box said to have been invented by
Athanasius Kircher (1601–1680) in the middle of the
seventeenth century, but it was not taken up with any
degree of popularity, notably in Germany and England,
until the end of the eighteenth century. This string
instrument, on which the wind produces sounds that
are considered musical, takes several forms; most often,
it is a box with a resonance chamber over which strings
of different kinds of materials are stretched all along
its length. They can be adjusted to produce this or that
sound, but, in view of the variations in the speed of the
wind, the frequency of the vibrations varies and so does
the sound that it produces. In a word, the wind becomes
the instrumentalist. Most often, the Aeolian harp is

placed on an open window. It is a domestic practice that funnels a collective sensibility.

For our purposes, the most important thing is that the Aeolian harp is evoked in the title of numerous scores, such as the first movement of the étude by Frédéric Chopin (1810–1849), opus 25, no. 1, and it was celebrated in literature several times, most notably by Samuel Taylor Coleridge (1772–1834).[2] It also came to designate all tones produced by the wind, both in the forests and on the moors, even when men did not intervene. Since antiquity, the wind god Aeolus has been attributed the characteristics of nature's musician, using trees as his instruments. At the end of the eighteenth century, few poets had forgotten about this music of the winds, whether delicate or savage. "Nature as the universal lyre – i.e. the Aeolian harp – is eminently romantic," wrote the philosopher Pauline Nadrigny recently,[3] a theme that was originally explored by Novalis (1772–1801). In 1822 Johann Wolfgang von Goethe (1749–1832) composed a poem initially entitled "He, She," but responding to the fashion of the day, and wishing to place it among those consecrated to nature's voice, he changed the title to "Äolsharfen," even though it was not about the wind or the harp.[4] In March 1811 (or 1812) Maine de Biran (1766–1824) chose this theme to define his own sensibility: "My imagination and my sensibility, raised to a high level, was like the Aeolian harps whose strings vibrated at the least breath of wind and made the most harmonious sounds."[5]

Much later, during the second half of the nineteenth century, the American transcendentalists Ralph Waldo Emerson (1803–1882) and Henry David Thoreau (1817–1862) took up this theme. On this point, the latter also referred back to Novalis and claimed that he himself was very sensitive to the "universal lyre." In July 1851, for example, Thoreau made an allusion in his *Journal* to the subtle music of the "Aeolian harp." He returned to this theme several times, notably referring to the sound produced by telegraph wires, which corresponded to his sensitivity to everything that came out of the air.[6] A little later, Eugène Delacroix (1798–1863), after having devoted a long passage in his *Journal* to Joseph Joubert (1754–1824), cited a quotation attributed to the latter, which was found in a manuscript discovered after his death: "I am like an Aeolian harp that emits some beautiful sounds but plays no melody."[7]

Let us return to the manifestations of the intimate union between sensibility and wind. Already in the seventeenth century – but this case is totally exceptional for its era – Mme de Sévigné (1626–1696) grew anxious in the face of the wind and the rain. She worried that her daughter, Mme Grignan (1646–1705), might find herself exposed to these "cruel winds from the south that knock over everything and could kill you." She detested certain strong and cold winds that blew in Brittany. They affected her health, she wrote on July 13, 1689, and, above all, they made her "melancholic."[8]

Readers of *Julie, or The New Eloise* will remember the importance that Jean-Jacques Rousseau (1712–1778) placed on the "desiccating" wind that, in the opinion of his hero Saint-Preux, killed nature. It concerned a diurnal thermal wind, characteristic of the Leman, the region around Lake Geneva, which blew from the east to the northeast.[9] However, Anouchka Vasak notes that Rousseau created "a break between the meteorological reality – that is, its exterior form – and the consciousness that sketched out the neat contours of the subject." In this sense, the man who first conjured up "the barometer of the soul" did not push the aforementioned traits to their logical conclusions in order to arrive at a concept of the "meteorological self."[10]

That concept shows up throughout nearly all of Joubert's writings, to which we have already referred in regard to the Aeolian harp. This author dreamed about "writing in the air," "even in the sky." He paid special attention to the rain and sunny weather. He was particularly endowed with a barometric soul that transmitted into his writing "this disjointed, disconnected, and sporadic principle."[11] In other words, it followed the pattern of the wind.

Not long ago, I analyzed in detail the theological foundations of "the discovery of the seaside."[12] The principal architect of this social phenomenon, Dr Richard Russel (1687–1759), perceived the winds as subject to the guidance of natural theology. If the storm, through the turbulence of the water that it engendered, was intended to modify the air by purifying it or replacing it, the sea winds were created by God, not only to drive ships but, more importantly, to cleanse the waters. Thus, we let go of the ancient image of the sea, as well as the presence of Aeolus and the nymphs, concepts to which we shall return. All the same, the desire to go and examine the great classical texts in Italy, in person, was paramount among travelers of this era, and especially among the English, who had been undertaking the Grand Tour for 200 years. In terms of our subject, the wind, this was particularly true in regard to the storms of the *Aeneid*.

The emotions aroused by the wind were therefore determined by conflicting concepts: either natural theology or classical studies. Later on, in light of the ideal of the sublime, the importance of storms figured in the works of James Thomson (1700–1748) and James Macpherson (1736–1796). And let us add to this discussion the fact that, in the seventeenth century, Robert Burton (1572–1640) recommended open air as a remedy for melancholy. In the middle of the Enlightenment, it was believed that the sea could calm the anxieties of the elites. A visit to the seaside re-established a contact with nature, which remedied all the ills of civilization.[13]

What is the place of the wind in this set of processes? Of paramount importance is the fact that the beach must be clean and the quality of its air must be pure. This is what justified the vogue for Brighton, recommended by Dr Russel as the ideal place. "The cliffs," wrote Anthony Reilhan (1715–1776) in an ode that he composed in honor of Brighthelmston (Brighton), "defended by hills to the northward, which intercept the land breezes and prevent their bringing any quantity of matter with them when they blow . . . ; and the southeast and southwest winds, which blow from the sea . . . , must be assistant in blowing off any accession of solid matter that may arise from the town."[14] Throughout the years, this attention to the quality of the air and the winds grew, whereas the discourses on the merits of the waters declined. Good respiration was essential. That is why doctors ordered women to take short walks on the dunes, followed by a bath in the afternoon. At the same time, Swiss doctors and their followers prescribed an "air cure." All these recommendations were to promote the inhalation of good air.

In England, such advice was addressed to invalids, who were quite numerous among those who came to the seaside to seek treatment. Luckily for us, one of them, Baronet Richard Townley, published a journal that he kept during a year-long stay on the Isle of Man in order to restore his health. An expert at listening to his own body, he went there in 1789 in order to take advantage of the sea air. His attention was drawn, above all, to the quality of the air and the wind. Each day he tried very hard to describe, in great detail, the effects that the winds had on his senses and his soul. Thus, according to him, the winds could be "agreeable," "comfortable," "favorable," "unfavorable," "boisterous," and so on. Above all, he appreciated the "sea breezes." In his opinion, the most delightful part of the day was "the coming tide," often accompanied by a "fresh breeze." The range of his excursions, always on foot, was extensive. After getting up, he would go out "to meet the freshness of the morning breeze," which was "brought to the sandy shore by the approaching tide" (as was the case on August 4, 1789). He noted the effect of the wind on his breathing, and that is why, early in the morning, he was in the habit of opening his windows in the hopes of stimulating his appetite.[15] In all things, Baronet Townley liked freshness. At the end of his book, he included a hymn to the Isle of Man. This "invalid" found there a number of quiet creeks and pools, good for bathing, which were sheltered from the winds. These creeks and pools fulfilled a major therapeutic function, based on a sensual engagement of the entire being, attentive to fragrances, imperceptible murmurs, and the slightest blowing of the breezes.

The Aeolian Harp

Let us now consider in more detail the emotions of individuals who were sensitive to the weather, especially those who left behind traces of themselves – that is, who wrote autobiographies. Of course, Bernardin de Saint-Pierre (1737–1814) belongs to this category, and his reaction to the certainty of the distant origin of atmospheric events has already been mentioned. In his *Études de la Nature*, he described "the pleasures of bad weather." For example, when it rained in torrents, he wrote: "I hear the whistling of the wind, mingled with the clattering of the rain. These melancholic sounds, in the night time, throw me into a soft and profound sleep." Clarifying his emotions, he confided: "In bad weather, the sentiment of my human misery is quieted by seeing it rain, while I am under covers; by hearing the wind blow violently, while I am comfortably in bed. In this case, I enjoy a negative felicity." Added to this sensation, in his opinion, were "some other attributes of the Deity." "Everything delicious and transporting in our pleasure," he wrote, "arose from the sentiment of infinity," especially "the far-off murmurings of the wind." Further on, Bernardin de Saint-Pierre inserted this ideal into the daytime moments that other painters and poets have celebrated: "the sunrise, the sunset, the murmur of the winds, and the darkness of the night."[16]

I shall now return in more detail to other weather-sensitive individuals whom we know paid much attention to the wind. François-René de Chateaubriand (1768–1848) emphasized the influence of the wind on his life in his *Mémoirs d'outre-tombe* [Memoirs from Beyond the Tomb], as well as in his novel *René* and his travel stories. "I was educated as the companion of the winds and waves. One of the first pleasures that I enjoyed was combating the storms." During his childhood, in the evenings at Combourg, he often remembered hearing "the murmur of the wind." Of strolling with his sister Lucile at the age of seventeen, he said: "our principal recreation consisted of walking side by side on the Great Mall: in spring on a carpet of primroses; in autumn on beds of withered foliage; and in winter on a covering of snow." Further on, he wrote about his "Autumn Joy": "It gave me indescribable pleasure to see the return of the tempestuous season . . . A bluish mist rose from the paths of the forests . . . The moaning or plaintive music of the wind whispered through the withered mosses . . . The frosts, by rendering communication less easy, isolated the inhabitants of the country." Later, delirious in his room, he confided: "The blasts of the north wind were to my ears the sighs of voluptuous delight." Afterwards, when he found himself walking "into the wood," he marched toward the adventure, "embracing the winds that escaped from me."[17]

René is a novel. All the same, though, it could be considered an autobiography. Alone in Brittany, after returning from Italy, while Amélie – that is, Lucile – is in Paris, René confesses: "I embraced it [the ideal object of some future affection] in the winds," adding: "the sounds that render passion in the void of a solitary heart resemble the murmurs the winds and the rains produce in the silence of the desert." He walks with determined steps, "my face burning, the wind whistling through my hair," crying out: "Rise, swiftly longed-for storm!" Finally, his tears seem to him to be "less bitter when I shed them over the rocks and among the winds." It goes without saying that all of this is expressed in Ossianic poetry, to which we will return at length. The confrontation with the wind, like the music of the storm as well as tears, are stereotypes of Ossianic poetry. It is this that René states clearly when he speaks of the "months of storms": "Sometimes I wanted to be a warrior wandering among those winds, clouds, and phantoms."[18] There are just two more Frenchmen left to mention who had the most acute sensitivity to the weather: Maine de Biran and Maurice de Guérin (1810–1839). I have shown elsewhere the precise notations, day after day, that the former made in his *Journal* concerning the rain, the wind, and the fair weather.[19] In his work, always haunted by what he described as his "preoccupation," the state of the sky was decisive to his mood, his spirits, and his melancholy. Thus, on February 12, 1813, Maine de Biran wrote: "A beautiful day . . . , a warm wind from the south." And the next day: "Rain. Wind from the southwest (a storm). I got up in a state of anxiety and depression." But on the following day, February 14, he wrote: "Rain, wind, and mild weather." For the following three days, the presence of the wind also figured prominently in his *Journal*. He even recorded the temperature of the wind. Thus, on February 18, he noted: "A cold wind," and added, "Inner worries persist. I feel not quite myself."[20]

There are too many of these kinds of statements to set down here. Sometimes, the state of the weather was recorded very precisely. Thus, on April 14, 1818, when the sky was clear, he wrote: "A cold wind blew . . . The air lost its sweetness; it was no longer springtime. The change . . . influenced my mood. I was preoccupied, suffering, and little disposed to work." In the months of May and June, he noted often the presence of the wind, either accompanied by rain or by clear skies. Thus, on

June 16: "Changeable weather with a great wind. I was fairly indisposed all day long." The absence or the presence of the wind in the *Journal* determined, in part, his humor, his health, his degree of concentration, and his receptivity for reflection or meditation.[21]

Claude Reichler underscores the changeability, the irregularity, the inner fluctuations, and the unpredictability that characterized Maine de Biran.[22] And all of these traits were equally traits which belonged to the wind. To be precise: in this regard, the author of the *Journal* was a native of the Bergerac region and had its "medical constitution." Nevertheless, we know and understand the wind's importance in this type of literature.

If the preceding shows Maine de Biran's sensitivity to the nature of the winds, this is explained only in an obscure manner. This sensitivity moves in a very different direction in the notes that Guérin made about the movements of the air. It seems to me that this young Romantic writer carried out the most profound analysis of the wind's effect on himself and others. That is why we will focus on him more than on Maine de Biran.

On May 23, 1833, in a letter written to his friend Raymond de Rivières, Guérin meticulously explained his sensitivity to the weather:

Unhappily, the state of my soul is subject to atmospheric influences, lightly administered, but, however light the effect, the actions of the atmosphere is not any less a burden for me, especially when the days are overcast and rainy. When the sky brightens up, I experience a lightening of the soul, and I feel a serenity and an impulse toward joy that prevails over the sadness of the darkness of the deepest kind.[23]

The wind – or, rather, the winds – "those fearful blasts from an unknown mouth" – and that is what concerns us here – determines his physical state. What Guérin considered "the voice of nature," he confided, "has acquired such empire over me that I rarely succeeded in freeing myself from the habitual preoccupation that she imposed on me." He followed this phrase with another passage that falls within the definition of the sublime: "To wake at midnight with the shrieks of the storm, to be assailed in the darkness by a savage and furious harmony, which subverts the peaceful sway of night, is something incomparable in the experience of strange impressions. It is a terrible delight."[24]

Guérin and one of his friends experienced "the violence of the wind." The former described what he defined as "wild struggles," which again falls within the notion of the sublime as it had been previously defined by Edmund Burke (1729–1797) and Emmanuel Kant (1724–1804). Guérin and his friend, these "souls of two beings five feet in height," were "planted upon the crest of a cliff [in Brittany], shaken like leaves by the violence of the wind." "We inclined our bodies and spread our legs apart to enlarge our mass so as to resist the wind with the greatest advantage, and our hands were attached to our hats to secure them to our heads." They lived, he explained, in "one of those mingled moments of sublime excitement and profound meditation, when the soul and nature, drawing together to their full height, confronted each other."[25]

We will not list Maurice de Guérin's every statement

devoted to the wind and to his intense and lifelong sensitivity to the weather. Let us cite one text, though, which is probably too long, but in which we hear with a particular intensity what the wind could arouse in this still very young man. It concerns a description of the weather on May 1, 1833:

> Heavens! How gloomy! wind, rain, cold ... Today I have seen nothing but showers streaming through the air in disordered ranks, driven furiously before a mad wind. I have heard nothing but this same wind wailing on every side of me with those pitiful and sinister wails which it catches or learns I know not where; one would say it seems the very blast of misfortune, of calamity, of all the afflictions which I imagine to be hovering in our atmosphere, shaking our dwellings and chanting its mournful prophecies about our windows. The wind, whatever it may be, at the same time that it was affecting my soul so sadly by its mysterious spirit, was agitating external nature by its material power, and perhaps also by something more; for who knows if we are acquainted with the whole range of the mutual relations and intercourse of the elements? I saw this wind from my window doing its utmost against the trees, driving them to despair ... In these days, there is revealed in the depth of my soul, in the innermost, the profoundest recesses of its being, a sort of despair altogether strange; it is like the abandonment of an outer darkness where God is not. My God! How comes it that my repose is affected by what passes in the air, and that the peace of my soul is thus given up to the caprice of the winds? ... I have become the sport of every breath that blows.[26]

The Aeolian Harp

I have come to the conclusion that Guérin's words, unjustly forgotten by the public at large, along with those of Victor Hugo (1802–1885), are the most fascinating ones that have been written about the wind, especially about its effects on nature and its repercussions on the soul. And let us not forget that, in these texts, an essential fact is very quickly revealed: the wind is an impenetrable enigma.

4

New Experiences of the Wind

On Sunday July 13, 1788, what was described as the greatest storm that had been unleashed on Western Europe in human memory spread in an unbroken path from Tours to Flanders. The storm moved with great speed from southwest to northeast. Appearing in Touraine at 6:30 in the morning, it reached Rambouillet at 8:00 and Pontoise at 8:30. Let us be clear: the scientific explanations of such a catastrophe were totally defective. This storm appeared as a genuine act of aggression.[1] Henri-Alexander Tessier (1741–1837), a member of the Royal Academy of Sciences, one of the three most important writers who drew up reports on this subject, considered the wind as the principal agent of the disaster.

> The wind, which seemed to follow a concerted plan, appeared as an invisible and malicious character. This agent drove, regulated, and directed everything. It swirled, it swept away the clouds, it twisted the trees, it blew from different directions ... It crossed deep valleys and high

hills, forests and rivers. In particular, it crossed the Loire and the Seine and contributed to the falling of hail in the countryside where it almost never fell. Everything was dug up, chopped up, smashed up, uprooted. Roofs were torn off, windows broken, cows and sheep killed or wounded. Fruit fell from the trees, and vegetables were shredded. Birds were killed and so were sheep.[2]

The "storm" left the countryside "ravaged" and "devastated." It seemed like a "real apocalypse." In particular, the wind's "aggression" hit the chateaux hard: the richer the house, the more spectacular the destruction. This was the case with the Chateau de Rambouillet, which featured heavily in the witness statements. It was reported that 11,749 wall tiles were broken, that roof tiles and slate tiles were "crushed," and that roofing was "knocked into the walls."[3]

Of course, the tempest of July 13, 1788, was soon to be perceived, according to the symbols it contained, as a harbinger of the Revolution. It was conceived as a "stormy disruption that was born out of and extended through the effects of its own power."[4] The path of the Great Fear of 1789 seemed to be the same as this great storm, and so the two were associated with each other. Anouchka Vasak analyzes this event with a great deal of attention. According to her, the tempest had induced another understanding of space and had given rise to the thought that it was no longer a stable and solid thing. "The pulverization of the windows in the chateaux," she writes, "referred to an order of the world where philosophy and physics were beginning to understand the real by recognizing its essential disorder."[5] In my opinion, the role of the "unleashed wind," its aggression on the countryside and its people, takes on a significance much greater, real, symbolic, and metaphoric than a storm at sea, which was then unjustly favored by painters.

The second half of the eighteenth century was literally sprinkled with great sea voyages. It was the age of the sailing vessel, which was designed to capture the wind. For this reason alone, these adventures should be situated at the heart of this book, but this will not be the case. The accounts of these voyages by the great navigators, such as Louis Antoine de Bougainville (1729–1811) and James Cook (1728–1779), made abundant references to the winds and the navigation techniques that they necessitated, but these texts showed that the winds were gravely misunderstood and that familiarity with them was close to zero, apart from the trade winds and several regional winds. Nothing – or almost nothing – was said about a desire to understand them, except for the use of instruments to determine their direction and to measure their velocity. The reading of maps at that time had as its sole goal the detection of shelters in which to take refuge in case of danger.

This situation was produced by a specific experience: it was the wind that regulated all the routes and made them possible or impossible. Sailors were completely subservient to the winds, which decided the time for casting off, among other things. Memories of their nature and the risks that they could produce weighed heavily on the choice of the route to follow. In the course of a voyage, a "gust of wind" – and, even more so, a hurricane – was to be feared. A "cross wind" could hamper movement, and so could a "standing breeze," which would require tacking. On the other hand, navigators appreciated it when the wind was calm, when it "gave way," when there was a "fair wind" or a "favorable wind," when it was possible to "run with the wind."

During the sailing, the winds required extremely complex operations that had, as their goal, the adaptation of the sail to suit the wind, but, for all that, innovation was limited, associated with only technical progress in the construction of vessels and their sails. If we read the navigators' reports from this era, in contrast to what is stated in literary works that transcribe the wind's imagery – and we will come back to this theme – the only notable change is the fact that the navigators abandoned the use of the Aeolian names of antiquity: they no longer talked about Boreas, Zephyrus, Eurus, and Notus. As a result, readers of these navigation reports find themselves obligated to endure tedious pages devoted to stories of repetitive operations. It is, therefore, not here where we will find new experiences of the wind. Throughout the entire length of the nineteenth century, the progress of the clippers was certain and obvious, but, paradoxically, it did not enrich our understanding of the wind.

Nevertheless, let us turn to Bougainville and his description of how he reached the Straits of Magellan in 1767 and 1768. Nothing is left out, from his dependence on the wind to how it imposed its rhythm on the sailors. On December 2, 1767, he wrote:

From the 2nd of December in the afternoon, when we got sight of Cape Virgins, and soon after of Tierra del Fuego, the contrary wind and the stormy weather opposed us for several days together. We plied to windward the 3rd till six in the evening, when the winds, becoming more favourable, permitted our bearing away for the entrance of Magellan's Straits ... At half past seven, the wind became quite calm and the coast covered with fogs; at ten it blew fresh again, and we passed the night by plying to windward. The 4th, at three o'clock in the morning, we made for the land with a good northern breeze.[6]

A month later, a hurricane hit:

In the night between the 21st and 22nd [January 1768], there was a calm interval. It seemed that the wind afforded us that momentary repose, only to fall harder upon us afterwards. A dreadful hurricane came suddenly from S.S.W. and blew with such fury as to astonish the oldest sailors.[7]

Let us emphasize this reference to memory:

Both our ships had their anchors come home and were obliged *to let go their sheet-anchor, lower the lower-yards, and hand the top masts. Our mizzen was carried away in the brails.* Happily, this hurricane did not last long. On the 24th, the storm abated.[8]

Bougainville concluded from these episodes that, all things considered, it was better to take the Straits of Magellan around South America rather than Cape Horn.

The Balloon "at the Head of the Wind"

At first glance, when air had become all the rage at the end of the eighteenth century, for all sorts of reasons cited above, we would have expected mainly balloonists to be leading this new fascination with the wind. In reality, history unfolded a little differently.[9] Certainly – and this was long the case until the invention of the dirigible – balloons were driven by the wind. These, as the saying went, were driven forward "at the wind's fancy." The balloonists experienced the winds as a decisive factor in their voyages, more so than those who navigated over the sea. Nevertheless, in reading the reports of these first adventurers in the sky, the fact of moving about in this ocean of air was the first thing to note – something that was expressed and described by Alexander von Humboldt, but only partially, and something that was, until then, totally unknown. Since the dawn of time, Icarus' dream had never been realized, but this navigation in the air, totally new in the eighteenth century, drew attention to the presence and the analysis of the wind as a vehicle for emotion.

Moreover, at the end of the Enlightenment, the void ceased to be automatically perceived as negative. It became possible to study it on its own terms and to think that it could cure the individual of too much civilization. With this perspective in mind, a trip through the air was a "confrontation with the void, with its dynamic emptiness." The attention given to the wind was similar to that given to the void. And when the sight of the earth disappeared, the trip through the air became a trip without landmarks. However, all trips, whether over land or sea, were guided by markers, especially those that were subject to the winds. Finally, the formless character of the sky rendered the balloonist unconscious of movement and caused him "to feel lost in its total immensity." The sensation of "unlimited fluidity" caused him to neglect the movement's motor – that is

---

to say, the wind. Moreover, in the opinion of the first balloonists, their flights, with few exceptions, were a rupture with the divine concept of the heavens.[10]

Not to be forgotten is the implication of religion and science in this voyage through space. For example, it was the balloonists who said that they felt like beings between heaven and earth, lost in the total immateriality of the experience. They felt as if the balloon allowed "bodily access to infinity."[11] That said, the balloon flight confirmed the non-existence of a celestial vault between the earth and the rest of the universe. This explains the emotions reported after the first flights in balloons by those who found themselves enfolded in them. More than the impression of feeling the wind in a new way, it was the sight of the earth and what could be discovered below that fascinated the first balloonists. They were the spectators of a vast panorama. They were said to have been struck by the solitude of the skies, the silence of the nights, and the noiseless movement.

Other important emotions expressed by the balloonists concerned the internal sensations. At first, there was a kind of euphoria, which was later attributed to the lack of oxygen in the air. The flight of the balloon, an escape from earthly smells, led to a different way of breathing, more in tune with the universe – or so it was believed. Let us not forget the supposed therapeutic value of the air and of altitude, because, it was thought, air gained in purity the higher one went. Moreover, the resistance to cold as claimed by the balloonist was explained by the fact that, in the balloon, they moved with the winds, whereas, on a high mountain, more often than not, mountaineers confronted the winds. Another conviction was put forward: the freshness and the quality of the air was assured by the fact that the balloon was always in motion, exercising an invigorating action, especially the perceived sensation it produced on the skin. In brief, in their reports, the balloonists insisted on the moral and physical purity of the skies. They maintained that this ocean of air, constantly on the move, purified the body and the soul.

Nevertheless, the ascent was not always without physical problems. The first balloonists experienced light-headedness and vertigo and had difficulty breathing above a certain altitude. In the middle of the nineteenth century, Paul Bert (1833–1886) went on to explain all the nefarious effects of altitude. He proved that they were the result of a lack of oxygen. For the first balloonists attained very high altitudes from very early on. As early as December 1, 1783, a balloon reached 9,000 feet. In 1785, another exceeded 15,000 feet. These achievements satisfied the obsession with heights and fulfilled the feeling that the balloonist had conquered the air in defiance of the elements and in confrontation with the void. Lost in this ocean of air, without landmarks, the balloonist allowed himself to be dominated completely by the air. He felt adrift, at the mercy of the winds, from which came an impression of tranquillity and peacefulness, accompanied by a silence that was at times perceived as holy.[12]

In France, the publication in 1872 of *L'Atmosphere: description des grands phenomènes de la nature* [The atmosphere: a description of nature's great phenomena] by Camille Flammarion (1842–1925) marked a turning point. Thanks to this widely disseminated work, the public at large learned about the sensations experienced by balloonists. Their testimonies in the second half of the century were generally both more precise and more elegant than those of the pioneers. Moreover, many more people began to have their own experiences of this kind.[13]

In the course of 1887 and 1888, Guy de Maupassant (1850–1893) embarked on two trips in a balloon called *Le Horla*. He related what happened in what he called his "aerial expeditions," the first one toward Belgium and the second one in the vicinity of Paris. What do they hold of interest for us? In confronting a scornful Émile Verhaeren (1855–1916), in whose opinion, to travel in a "flying basket" made a person a "slave to the wind," Maupassant was reduced to cryptic phrases. In the course of one of these balloon trips, he wrote: "You feel nothing, you float, you rise, you fly, you coast." To begin with, he said nothing about the impressions aroused by the wind. Still, fulfilling here the role of a reporter, he felt compelled to describe the full range of his emotions: "a deep and unknown feeling of being," "an infinite rest of oblivion and indifference to all things," a movement "without noise, without troubles, and without worries."[14]

That said, this habitual sailor of rivers large and small could not totally neglect the wind. Concerning the latter, he delivered some trenchant remarks. Of the balloon, he wrote: "An enormous toy, free and docile, that obeys with an amazing sensibility, but it is also, and above all else, a slave to the wind, over which we have no command." Then, without forgetting about the wind, Maupassant became excited: "Deliciously inert, we travel though space. The air is brought to us and makes us beings that resemble it, beings that are mute, joyful, and crazy." And he sums up his experiences: "We no longer have any regrets, any projects, or any hopes."[15]

Upon his return, Maupassant came back to the wind. As it descended to earth, the speed of the balloon could be measured; it went as fast as the wind; and the wind became perceptible. "And I heard presently," he wrote, "leaning over the basket, the great noise of the wind in the trees and through the crops ... I said to Captain Jovis [who was leading the adventure]: 'Oh! How it blows!'" Having been contradicted, he went on: "I insist: it is the wind! I am sure because my ears know it so well, having heard it so often whistling through the ropes."[16]

This testimony about experiencing the wind from inside a balloon confirms what we have already said: the wind is the essential motor of the apparatus, but in a silence, a peacefulness, a fluidity in motion, and a purity that tends to make us forget about it, until the closeness of the earth makes us once again hear its noisy and menacing bellowing, which seems to announce, as it does on the sea when it is unleashed, the danger of ruin and death.

The Sandstorm in the Desert

The sandstorm in the desert has been known since antiquity. Herodotus once alluded to it. In 1730 James Thomson (1700–1748) devoted a gripping passage to it in his *Seasons* in his description of summer. However, in the nineteenth century, for the first time, a number of explorers related these kinds of gripping experiences. North Africa, especially the Sahara and Egypt, was the usual place for these encounters.

René Caillié (1799–1838) in the course of his famous trip to Timbuctoo was subjected to "sandstorms" on several occasions. On May 23, 1828, the wind from the east blew with a great deal of violence. He wrote:

> It threatened to bury us under the mountains of sand which it raised ... What distressed us most during this horrible day was the pillars of sand ... One of the largest of these pillars crossed our camp, overset all the tents, and whirling us about like straws, threw us one upon another in the utmost confusion. We knew not where we were and could distinguish nothing at the distance of a foot. The sand wrapped us in darkness like a thick fog, and heaven and earth seemed confounded and blended into one.[17]

During this natural disaster, the dismay was widespread. Plaintive cries were heard from every side. Many of the men cried out to God, pleading with all their might.

> 'There is no God but Allah, and Mahomet is his Prophet!' Through these shouts and prayers, and the roaring of the wind, I could distinguish, at intervals, the low plaintive moans of the camels ... Whilst this frightful tempest lasted, we remained stretched on the ground, motionless, dying of thirst, burned by the heat of the sand, and buffeted by the sun, whose disk, almost concealed by the cloud of sand, appeared dim and shorn of its beams.[18]

This "terrible hurricane" lasted three hours.

This description, worthy of the most talented Orientalist painters, was not surpassed by many other travelers' stories of sandstorms. In the midst of this menacing disaster by and in the sand, we find the same human reactions that are often described during maritime shipwrecks, except that, in this case, the wind is the only author of the drama.

The Sahara was thus the territory that expressed

the power of Allah. As the scholar Guy Barthélemy remarked, the setting of the sandstorm was an event heightened by its scenery, by "the extreme nature of the desert geology," as emphasized later by Pierre Loti (1850–1923). Both the sandstorm and the desert were substantial evidence of the transcendence that was the theme of the Orientalists. With the *vastitas* – the desert's emptiness, silence, and strangeness – it had become, thanks to the wind, the privileged place for listening to God's voice in a setting that evoked immensity and eternity. The sandstorm, Barthélemy wrote, "illustrated the ambiguity of the desert, which is the same as that of God. When it blows, it speaks to the soul." "When this puff of wind becomes a storm through its terrifying intensification, the traveler runs a risk comparable to that when a man wants to contemplate the face of God."[19]

This metaphysical interpretation of the sandstorm unleashed as a mysterious desire was found in music, as witnessed in 1844 in the symphonic ode by Félicien David (1810–1876) entitled *Le Desert*.

Lower your heads!
The simoom, the wind of fire
Passes, like a plague from God.
Allah! Have pity on your believers!
Allah! Remember those with fervent hearts!
Heaven is no longer – Oh, Allah! Oh, Allah!
Hell presses upon us![20]

In the numerous stories of sandstorms left by writers of the nineteenth century, at least in the eyes of French readers, the khamsin (also known as a simoom), a great vertical cloud, a whirlwind, that rises up in Egypt in the desert of Kossier, takes pride of place. The account that Gustave Flaubert (1821–1880) gave of it in 1850 is a kind of "sensory journal."[21]

> It is hot; on the right a khamsin dust-cloud is moving our way from the direction of the Nile (of which all that we can faintly see now is a few of the palms that line the bank). The dust-cloud grows and comes straight at us – it is like an immense vertical cloud that before enveloping us is already high above us for some time, while its base, to the right, is still distant. It is reddish brown and pale red; now we are in the midst of it . . . I feel something like terror and furious admiration creep along my spine; I laugh nervously; I must have been pale, and my enjoyment of the moment was intense. As the caravan passed, it seemed to me that the camels were not touching the ground, that they were breasting ahead with a ship-like movement . . . [22]

A little later on, he continued:

> The hot wind comes from the south; the sun looks like a tarnished silver plate; a second dust-spout comes on us. This one advances like the smoke from a conflagration, suet-colored, with jet-black tones at the base. It comes . . . and comes . . . and the curtain is on us, bulging out in volutes below, with deep black fringes. We are enveloped by it: the force of the wind is such that we have to clutch our saddles to stay on. When the worst of the storm has passed, there comes a hail of small pebbles carried by the wind: the camels turn their tails to it, stop, and lie down.[23]

There is no question here of metaphysical dramatization. In Flaubert's experience, what was at play was something closer to the sublime. The unleashed whirlwind and its sensation of a great vertical cloud was, for the traveler, the occasion of experiencing a pleasure that was both incredible and inexpressible. Nevertheless, contrary to the encampment devastated by the storm as described by Caillié, Flaubert's notebook related the experience of the sandstorm by a caravan on camel back. It incites notions of a kinetic nature.

In the middle of the nineteenth century, the public's knowledge of the simoom was spread largely through the publication of *Five Weeks in a Balloon* by Jules Verne (1828–1905). Its success was enormous. Thus, in Limousin school libraries, according to a report by the teachers of La Creuse in 1877, it was the most popular book for readers. The simoom, which was seen and experienced twice by the balloonists, was described in detail. Both episodes were determined by changes in the direction and strength of the wind. In the first episode, below the balloonists:

> The plain was agitated like the sea shaken by the fury of a tempest; billows of sand went tossing over each other amid blinding clouds of dust; an immense pillar was seen whirling toward them through the air from the southeast; the sun was disappearing behind an opaque veil of cloud whose enormous barrier extended clear to the horizon, while the grains of sand went gliding together with all the supple ease of liquid particles, and the rising dust-tide gained more and more with every second. Ferguson's eyes gleamed with a ray of energetic hope. 'The simoom!' he exclaimed.[24]

After the navigators had jettisoned the ballast, the balloon rose up above the simoom and found itself carried "along with incalculable rapidity away above this foaming sea."[25] Verne had evidently done his homework by reading travelogues and had thus introduced a multitude of readers to the sandstorm.

The desert was not the only stage for the union of wind and sand. Henry David Thoreau (1817–1862) described several episodes of wild meteorological events during his trip to Cape Cod. On this small spit of land, according to him, the wind was omnipresent. It stripped the houses of their surroundings; it threw up sand onto the rocks; it whipped up the waves like sails. Its presence was particularly intense at Provincetown, especially in the section known as "the desert." Thoreau described one of his own experiences:

> The wind was not a Sirocco or Simoom, such as we associate with the desert, but a New England northeaster ... And we saw what it must be to face it when the weather was drier, and if possible, windier still – to face a migrating sand-bar in the air ... to be whipped with a cat, not o' nine-tails, but of a myriad of tails, and each one a sting to it.[26]

The apparent attraction that Cape Cod exercised on Thoreau was the result, primarily, of the wind and the sea and, secondarily, of the rocks and the sand. In this, the author transcribed the taste largely taken up in the West in the middle of the century, when the last fires of Romanticism still reigned, especially in the United States, under the name of Transcendentalism, whose high priest was Ralph Waldo Emerson (1803–1882).

The Wind in the Sequoias

During the second half of the eighteenth century, a new experience developed. Barbara Stafford called it *Voyage into Substance*.[27] It took place in a virgin and exotic natural landscape that was astonishing because of its vegetation. There, in a tropical rainforest, it aroused new emotions. This new expansive experience of exoticism in nature was described by Alexander von Humboldt (1769–1859) and Charles Darwin (1809–1882), among other scientists, and it continued a little later, at the end of the nineteenth century, with expeditions – "surveys" – in the American Far West. It was in this context, as the great national parks emerged, that those whom we could consider the successors of Transcendentalism – and the pioneers of the ecological movement – set out to listen to the wind in the forests of the giant trees. John Muir (1838–1914) was thus the greatest explorer of the winds – of their origin, their movement, and, above all, their words and their scents in the forests of Yosemite. He pushed to the extreme the attempts at listening that we have considered here. This adventurer, with his insurmountable passion, was thus ready to produce a history of the emotions aroused by the wind.

For five winters between 1868 and 1872, he set himself up in Yosemite Valley. His goal was to "*look at the winds*."

> Most people like to look at mountain rivers and bear them in mind, but few care to look at the winds, though far more beautiful and sublime, and though they became at times about as visible as flowing waters. When the north winds in winter are making upward sweeps over the curving summits of the High Sierras, the fact is sometimes published with flying snow-banners a mile long. Those portions of the winds thus embodied can scarce be wholly invisible, even to the darkest imagination.[28]

The wind's voice was integral to a series of sounds in nature, particularly under certain circumstances. Thus, in autumn, the winds' sigh was even sweeter than "the gentle ah-ah-ing filling the sky with a fine universal mist of music." However, "winter comes suddenly, arrayed in storms ... You hear strange whisperings among the tree-tops, as if the giants were taking counsel together."[29] Then the winds unleashed constituted the storm's splendor. In particular, the latter resulted from gusts of wind, as described by Muir.

In Utah, he was subjected to one of the largest storms that he had ever witnessed, punctuated by these "gusts of winds." One day at 4:30 in the afternoon, a dark brown cloud appeared:

> In a few minutes, it came sweeping over the valley in a wild roar, a torrent of wind thick with sand and dust, advancing with a most majestic front, rolling and overcoming like a gigantic sea-wave ... The bending trees, the dust streamers, and the cold onrush of everything movable [gave] it an appreciable visibility that rendered it grand and inspiring.[30]

In Muir's opinion, the storm winds distributed "their benefactions in the most cordial and harmonious storm-measures."[31]

The greatest pleasure that he experienced was in noticing what united the wind and the forest. He detailed it according to the essences of the trees.

> The mountain winds, like the dew and rain, sunshine and snow, are measured and bestowed with love on the forests to develop their strength and beauty. However restricted the scope of other forest influences, that of the winds is

---

universal . . . The winds go to every tree, fingering every
leaf and branch and furrowed bole; not one is forgotten
. . . They seek and find them all, caressing them tenderly,
bending them in lusty exercise, stimulating their growth,
plucking off a leaf or a limb as required.[32]

All commentary here would be superfluous, but
Muir's enthusiasm was not limited to just these details.
"There is always something deeply exciting, not only
in the sounds of the wind in the woods . . . but in their
varied waterlike flow as manifested by the movements
of the trees." And these movements are determined by
the trees' essence. In effect, Muir suggested that "the
gestures of the various trees made a delightful study."[33]
For example, while hiking through the Sierras in 1874,
he noted:

> Young sugar pines, light and feathery as squirrel-tails,
> were blowing almost to the ground, while the grand
> old patriarchs, whose massive boles had been tried in a
> hundred storms, waved solemnly above them, their long,
> arching branches streaming fluently on the gale, and every
> needle thrilling and ringing and shedding off keen lances
> like a diamond.[34]

There then followed a characterization of the way that
these Douglas firs, silver pines, and other evergreens
behaved in the wind. Muir allowed himself to be swept
along, he assured us, by "this passionate music and
motion." As a specialist, he emphasized that each tree
"was expressing itself in its own way."[35]
The religious spirit that impregnates Muir's sight and
sound is revealed particularly in what he wrote about
the silver pines blowing in the wind:

Colossal spires of 200 feet in height waved like supple goldenrods chanting and bowing low as if in worship ... The force of the gale was such that the most steadfast monarch of them all rocked down to its roots with a motion plainly perceptible when one leaned against it. Nature was holding high festival, and every fiber of the most rigid giants thrilled with glad excitement.[36]

The individual emotions of the enormous trees were not limited to the responses brought by the wind's assaults. These attacks transported a music and a range of scents into the area. From the top of his perch, to which we will return, Muir tried "to feast quietly on the delicious fragrances that was streaming past." "But from the chafing of resiny branches against each other and the incessant attrition of myriads of needles, the gale was spiced to a very tonic degree. And besides the fragrance from these local sources, there were traces of scents brought from afar." In effect, the wind that blew came from the sea, where it rubbed "against its fresh, briny waves," before it dodged in and out of the sequoias, insinuating itself in "rich ferny gulches" and dispersing in "broad undulating currents over many a flower-enameled ridge." In brief, "the winds are advertisements of all they touch, however much or little we may be able to read them, telling their wanderings even by the scents alone."[37] Thus, in finishing up, Muir recalled the phrase about the smells carried on the wind to the sailors as they approached land.

This explorer's experience of the forest, as far as it concerns us, culminates in a climbing trip that he made in the Sierras among the Douglas spruces 200 feet tall. His goal was to enjoy the particular music of the winds, which was perceptible in what he called "his lofty perch."[38] He wanted to perceive and to taste the Aeolian music played by the pine needles high up in the air. Muir enlarged on the sounds of this music in the forest:

> The profound bass of the naked branches and boles booming like waterfalls; the quick tense vibrations of the pine-needles, now rising to a shrill, whistling, hissing, now falling to a silky murmur; the rustling of laurel groves in the dells; and the keen metallic click of leaf on leaf – all this was heard in easy analysis when the attention was calmly bent.[39]

At the end of two hours spent in his "lofty perch," Muir was convinced that, within nature, there was no manifestation of danger or any sign of reprobation on the part of the trees. The wind in the forest arouses an invincible elation as far from exultation as from fear. This experience led this hero of the forest to argue against Darwin's theories and to refuse to accept any idea about "the universal struggle for existence."[40]

In order to understand better the thoughts but, above all, the emotions felt by Muir, such as he described them, it is necessary to take into account a jolt to the memory that he had experienced. His journeys to Yosemite and elsewhere were a strong reminder of his childhood in Scotland, which he recounted much later, in 1913, in *The Story of My Boyhood and Youth*.[41] One day, on an excursion in Florida, a sea breeze filtered across the palmettoes and the vine tangles, and he confided that it "awakened and set free a thousand dormant associations, and made me a boy again in Scotland, as if all the intervening years had been annihilated."[42]

I am confident that it is appropriate to add John Muir to the long list of writers who, long before the memory of the madeleine that brought back so much to Marcel Proust, spoke of a vivid experience which abruptly merged the past and present, whatever the moment.

5

The Tenacity of the Aeolian Imagination in the Bible

Throughout the centuries, and above all since the Renaissance, explanations of the wind which drew on our imagination compensated for our poor factual understanding of it. In the West, the foundation of this imagination came from certain texts from the Bible and Greco-Roman mythology. Subsequently, these elements have nourished the principal epics that peppered the history of literature. It would be bad form to neglect this collection of foundational literary works. At the same time, it is clear that fervent admiration of epic poetry has faded since the middle of the twentieth century, both in secondary education and in the works read by the educated public. Nevertheless, the biggest risk a historian can take is to appear psychologically anachronistic. Emphasizing the primary importance of Greco-Roman or biblical epics could put a smile on the face of today's critics, but, for those who are interested in the history of the wind, to ignore Homer and Virgil or Milton and Klopstock, or Tasso, Camões, and Ronsard, or, later still, Macpherson and Thomson would be to misunderstand everything about what has guided the Aeolian imagery during the last few centuries. Understanding this imagery is what will constitute our goal in the following chapters.

It is very difficult to deal with the presence of the wind in the Bible, particularly in the Old Testament, where everything is very subtle because the narrative is often confused with the presence or the breath of God. However, the passages concerning the wind are innumerable. Let us consider them methodically.[1]

According to Genesis, before the creation of the world and that of light, "a mighty wind swept over the surface of the waters."[2] Such is the prehistory of the wind in the Bible. Later, at the end of the Flood, "God made a wind pass over the earth, and the waters began to subside."[3] Subsequently, it is the storm cloud that takes precedence, and we must not confuse it with the wind. So it was when Yahweh met Moses on Mount Sinai. What unfolded between God and Elijah in the first book of Kings is more difficult to interpret. God ordered the prophet to go up the mountain: "For the Lord was passing by: a great and strong wind came rending the mountains and shattering the rocks before him, but the Lord was not in the wind."[4] Nor was he in the earthquake or the fire, both of which preceded "a low murmuring sound"[5] that signified the presence of Yahweh. And this gave rise to the thought among the faithful that it was not in the din and violence of the world where God was present.

We read in the Book of Job a portrait of God's power. Allusions are made to the rain, fog, and clouds, as well as to lightning and thunder, but we do not find many allusions to the wind per se. Still we read that "the east wind lifts him up and he is gone; it whirls him far from home."[6] Furthermore, it is acknowledged that "wicked men . . . are like chaff driven by the wind,"[7] but, for the pious, God makes "a sanctuary from wind and storm."[8]

The author (or authors) of the Psalms is more loquacious. To read these poetic prayers is to discover a series of allusions. In Psalm 18, we read that Yahweh flew down "on the wings of the wind,"[9] and the author writes that, at his coming, "the foundations of the earth were laid bare . . . at the blast of the breath of his nostrils."[10] Here, again, we perceive the wind as God's breath, similar to the passage in Genesis where it is the precursor of the Creation. In the same way, in Psalm 50, it is written that God is coming: "A consuming fire [or raging wind] runs before him and wreaths him closely around."[11] Until then, the wind appeared only when the presence of God was affirmed. It is expressed differently in Psalm 77 when referring back to the crossing in the desert. This time, as it was at the end of the Flood, the wind is the instrument of the Deity who commands it: "He lets loose the east wind from heaven and drove the south wind by his power."[12] And, in Psalm 104, we read: "[Thou] takest the clouds for thy chariot, riding on the wings of the wind, and makest the winds thy messengers."[13] Yahweh is celebrated as gentle and "great" in Psalm 135, where he "brings the wind out of his storehouses."[14] Among the objects that God created, the Psalmist cites "the gales of wind obeying his voice."[15]

Psalm 107 is of particular interest to us. It presents God as one who saves and delivers men, notably those "who go to sea in ships," from all perils that are unleashed by storms:

> At his command the storm-wind rose
> and lifted the waves high.
> Carried up to heaven, plunged down to the depths,
> tossed to and fro in peril,
> they reeled and staggered like drunken men,
> and their seamanship was all in vain.
> So they cried to the Lord in their trouble,
> and he brought them out of their distress.
> The storm sank to a murmur
> and the waves of the sea were stilled.[16]

This is a text that foreshadows a scene from the Gospels in which Jesus calms the waters.

Ecclesiastes in its prologue goes into ecstasy before the regularity of natural phenomena. Thus, "the wind blows south, the wind blows north, round and round it goes and returns full circle."[17] Here is a symbol of the vanity of the world: "They are all emptiness and chasing the wind."[18] Elsewhere, the wind's incomprehensibility symbolizes God's impenetrability. Subsequently, the wind's imperceptibility and its emptiness are emphasized several times as symbols of human failings. Thus, in the Wisdom of Solomon, "the swarming progeny of the wicked will come to no good . . . They will be shaken by the wind and by the violence of the winds uprooted."[19] And, again, "the hope of godless men" is swept away like "smoke which the wind whirls away."[20] They will be chastised, Solomon tells us: "A great tempest will arise against them and blow them away like chaff before a whirlwind."[21] More clearly, it is written in Ecclesiasticus:

> There are winds created to be agents of retribution,
> with great whips to give play to their fury;

on the day of reckoning, they exert their force
and give full vent to the anger of their Maker.[22]

Consequently, the wind as the instrument of God's punishment makes a return.

Throughout the Bible, of all the signs of God's presence, majesty, glory, and power, the wind is more insistently a manifestation of his punishment. "At his will, the south wind blows, the squall from the north, and the hurricane."[23]

Now let us turn to the books of the prophets. Jeremiah conjures up the wind in his vision of Jerusalem's fall as a result of Israel's transgressions.

A scorching wind from the high bare places in the
    wilderness
        sweeps down upon my people,
no breeze for winnowing or for cleansing;
a wind too strong for these
    will come at my bidding . . . [24]

Of God's anger – and let us emphasize his role in this catastrophe – Jeremiah writes:

At the thunder of his voice the waters in heaven are
    amazed;
he brings up the mist from the ends of the earth,
he opens rifts for the rain
and brings the wind out of his storehouses.[25]

A recurring idea appears here: there are four winds on earth, and each of them obeys God. Thus, we read in the oracle against Elam:

I will bring four winds against Elam
    from the four quarters of heaven;

I will scatter them before these four winds,
   and there shall be no nation
to which the exiles from Elam shall not come.[26]

From the time Ezekiel perceived God's chariot, he was enveloped in a wind from a storm blowing from the north – a most terrible wind. Against the false prophets, Yahweh declared: "In my rage, I will unleash a stormy wind."[27] In the Book of Daniel, on the other hand, it is a "moist wind"[28] that saves the three young men of God, and in their Song of the Three, they praised God, saying: "Bless the Lord, all winds that blow."[29] In the Book of Nahum, God makes the wind an instrument of his anger, and it is very violent. Thus, in the prayer entitled "The Vengeance of the Lord," it is written:

The Lord is a jealous god, a god of vengeance;
The Lord takes vengeance and is quick to anger.
In whirlwind and storm he goes his own way.[30]

In Zechariah's eighth vision, the cliché of the four winds – as in ancient mythology – imposes itself on the reader with great clarity and force. After having raised up four chariots, the angel of the Lord declares:

These are the four winds of heaven which have been attending the Lord of the whole earth, and they are now going forth. The chariot with the black horses is going to the land of the north, that with the white to the far west, that with the dappled to the south, and that with roan to the land of the east. They are eager to go and range over the whole earth.[31]

Zechariah goes on to say: "[The angel] called me to look and said: 'Those going to the land of the north have given my spirit rest in the land of the north.'"[32]

Consequently, there are many representations of the wind and the winds in the Old Testament and in the Apocrypha. Initially, it was a gentle exhalation which surrounded God and manifested his presence. Little by little, it became a sign and a witness to his power, majesty, and mercy. Finally, it was proclaimed an instrument of his anger, launched from all sides in the form of four winds unleashed on the earth.

Rarer, but still decisive, is the representation of the wind in the New Testament. It appears in the Gospels, the Acts of the Apostles, and Revelation. The episode of the storm calmed by Jesus in the Gospel of Matthew has been seized upon throughout the centuries. After the waters had become violent, the master was awakened by his disciples. "Then he stood up and *rebuked the wind* and the sea, and there was a dead calm." Seized with amazement, the disciples asked themselves: "What sort of man is this? Even the wind and the sea obey him."[33] Mark relates the same episode in his Gospel, but he does so with more detail. He describes the coming of "a heavy squall," and then Jesus, upon waking, "rebuked the wind and said to the sea: 'Hush! Be still!' The wind dropped and there was a dead calm."[34] In another episode, as related by Matthew, Peter tried to join Jesus, who was walking on the water. He took a few steps, but "when he saw the strength of the gale, he was seized with fear." Jesus saved him, and "they then climbed into the boat, and the wind dropped."[35]

In other words, in these episodes Jesus demonstrates that he has a power equivalent to that of the God of the Old Testament. This is confirmed when he announces the manifest glory of the Son of Man: "He will send out angels and gather his chosen from the four winds, from the farthest bounds of the earth to the farthest bounds of heaven."[36] Once again, the four winds represent the earth in its totality. As for the Gospel of John, he perceived the "strong wind" that blew upon the lake as the announcement of the coming of the Holy Spirit.[37]

The most important scene in the Acts of the Apostles, as is well known, is that of Pentecost, when the disciples were all gathered together after Jesus' death. It is usual to privilege the coming of the Holy Spirit in the form of "tongues like flames of fire," but it is also written: "When suddenly there came from the sky a noise like that of a strong driving wind, which filled the whole house where they were sitting."[38] These passages are reminiscent of several manifestations of the divine as are contained in the Old Testament.

Paul's trip to Rome is also a very famous story, especially the narrative about the risk of shipwreck near the coast of Malta, an event that was caused by "a fierce wind, the 'Northeaster' as they call it, [which] tore down from the landward side."[39] It has often been assumed that navigation on the Mediterranean, the theological sea, was the prefiguration of the entire course of human life from danger to safety. It is with this perspective in mind that it is advisable to interpret the wind that blows here as a symbol of life's troubles, which the sinner must overcome.

Finally, in Revelation, we find once again the presence of the four winds emanating from the four corners of the earth: "After this I saw four angels stationed at the four corners of the earth, holding back the four winds so that no wind should blow on sea or land or on any tree."[40] Briefly, then, in some kind of geographic or cosmic form, the value of the winds is reiterated here.

Paying attention to the Bible is indispensable in a modern history of the wind's imagery. It should be noted that, since the invention of the printing press, the Bible was the most widely read book in the West, and the Christian liturgy played a large part in this. Among Protestants, it was found in the most humble of homes. Without the reading of the Psalms and other books in the Bible, especially in England, natural theology, which had done so much to celebrate the contemplation of nature, would not have existed at all.

6

The Epic Power of the Wind

The reader of *The Odyssey* is kept constantly informed about the orientation of the winds; it is the winds that guide Ulysses' itinerary, causing constant detours. In no other masterpiece of epic literature pre-dating the nineteenth century do they play such a similar role, not even in *The Aeneid* or even in Camões' *Os Lusíadas* (1572). It is they, let us repeat, that drive the complications of these journeys, but they are in the hands of the gods: Zeus, Poseidon, Athena, and Aeolus, their master. There are four of them – as there are in the Bible – and each one possesses its own name: Boreas, Notus, Eurus, and Zephyrus. Each one harbors its own form of violence, its own kind of sensory characteristics, from auditory to tactile. They are, in short, beings – obviously obedient ones – each with its own personality. They are the instruments of the anger, jealousy, desire for vengeance, or forewarning fears of the gods.

The intense presence of the winds is explained in Book X of *The Odyssey*, when Ulysses and Aeolus, their master, still submissive to Zeus, confront each other. Ulysses disembarks and goes to Aeolus' house. The latter

> gave him a bag made of skin taken off a nine-year ox, stuffed full inside with the courses of all the blowing winds, for the son of Kronos [Zeus] had set him in charge over the winds, to hold them still or start them up at his pleasure. [Ulysses] stowed it away in the hollow ship, tied fast with a silver string, so there should be no wrong breath of wind, not even a little, but he set the West Wind [Zephyrus] free to blow him and carry the ships and the men aboard them on their way.[1]

Alas, on the tenth day, when the fatherland appeared on the horizon, Ulysses fell into an "unfortunate sleep," and the members of his crew, imaging that Aeolus had put gold and silver into the bag, took counsel with one another, and decided to see what was inside. "And they opened the bag and the winds burst out. Suddenly the storm caught them up and swept them over the waters, weeping, away from their own country."[2] It took them to Aeolus' island, and he abused them and chased them away. From that time onward, each of the four winds, which the author never neglects to name, is unleashed in its turn. They divide up their roles and antagonize the hearts of Ulysses and his companions. Sometimes, obeying the gods, they unite. In order to avoid citing too many of these incessant episodes, let us content ourselves in presenting just the one that unfolds in Book V.

Poseidon, angry against Ulysses for traveling on the Phoenician seas, takes his trident and unleashes the gales of the four winds.

East Wind [Eurus] and South Wind [Notus] clashed together, and the bitter blown West Wind [Zephyrus] and the North Wind [Boreas] born in the bright air rolled up a heavy sea . . . So the winds tossed her on the great sea, now here, now there, and now it would be South Wind [Notus] and North Wind [Boreas] that pushed her between them, and again East Wind [Eurus] and West Wind [Zephyrus] would burst in and follow.[3]

It was then that Athena, Zeus' daughter, "fastened down the courses of all the rest of the storm winds and told them all to go to sleep and to give over; but stirred a hastening North Wind [Boreas] and broke down the seas before him."[4]

Homer's words are gentler when Proteus speaks of "the limits of the earth where fair-haired Rhadamanthys is and where there is made the easiest life for mortals, for there is no snow, nor much winter there, nor is there ever rain, but always the stream of the Ocean sends up breezes of West Wind [Zephyrus] blowing briskly for the refreshment of mortals."[5]

We have not said everything there is to say about the winds in *The Odyssey* or in Greco-Roman mythology, but it is important to specify that Boreas (Aquilon in Latin) is the north wind; Zephyrus (Favonius in Latin) is the west wind; Notus (Amster in Latin) is the south wind, and Eurus (which has the same name in Latin) is the east wind.

Paradoxically, in *The Aeneid*, Virgil does not give to the winds as central a place as Homer does in *The Odyssey*. However, to judge by the epics of modern times, *The Aeneid* takes its place in the first rank of poetic references in regard to the actions of the winds and the storms. It was through this epic that, since the Renaissance, the wind's imagery has been magnified, exalted, and nourished. To neglect this literary genre, let us repeat, would be to miss the point of our project. Given the duality of the foundation that we have just analyzed, no doubt too briefly, we can divide epics, the historical sources of the imagery of the wind, into two categories: the first is made up of those under the influence of the biblical tradition and the second is linked to the influences of the Greco-Roman tradition. In the former are the poems of Guillaume Du Bartas (1533–1590) in the sixteenth century, John Milton (1608–1674) in the seventeenth century, and Friedrich Gottlieb Klopstock (1724–1803) in the eighteenth century; in the latter, *Gerusalemme liberata* [Jerusalem delivered] by Torquato Tasso (1544–1595), *La Franciade* by Pierre de Ronsard (1524–1585), and *Os Lusíadas* by Luis de Camões (1524–1580), even though Catholic missionary aims are clearly present in this last epic.

Du Bartas' *La Sepmaine* [The week], a poem dating from the end of the sixteenth century, tells the story of the Creation. In particular, it provides a description of Eden in an epic fashion. Today this work is totally forgotten, but it has its importance. The relevance to us occurs in Du Bartas' evocation of the Edenic wind. Above all, he devotes a long chapter to the second day, to the formation of heaven. He takes up the creation of the winds, which he characterizes as "windy elements." According to him, there exist two elementary spheres: "that of air and that of fire." But air is a zone that is "biblically non-existent." Du Bartas claims in an abashed manner that there is a middle zone, an intermediate space – "the sphere of troubled air" –

which is defined by its instability. It is there that the winds are situated, and he refers to this sphere as "the storehouse of the winds." In his opinion, it is a zone of "trouble and strife," symbolizing "the uncertain and perilous march of Christianity."[6]

More specifically, as far as the winds are concerned, after noting the effects of "their noisy crossings," Du Bartas distinguishes four times of day, four humors, four elements, and four ages. He then details the characteristics of two kinds of atmospheric conditions. The first is formed by the winds, which act most directly on the universe of living things and exercise a beneficial effect. This discussion prefigures what will be said in the Enlightenment: "The wind purges the polluted air; it causes fruit to grow; it fills the sails of ships; it turns the arms of windmills." However, it also has harmful effects: for example, it causes hail to fall, "which is an indication of the world's fragility and a symbol of a prefiguration of human disorders." As to the second kind, air is constantly changing its form and its role. This is where Du Bartas places the storms – and also the comets. It is made up of winds that could be interpreted as divine signs, as forewarnings, notably of God's anger, because of its closer proximity to heaven. In conclusion, in Du Bartas' opinion, the wind is deployed in a middle zone between heaven and earth, impenetrable to science.[7]

The greatest epic of modern times, if we stick to the extent of its distribution and its dedicated admirers, is none other than Milton's *Paradise Lost*. The wind is there, playing the role that it plays in the poems with Greco-Roman references, but it is also there, displaying the two aspects from the biblical tradition: a gentle

Edenic wind before the Fall and a terrible, violent wind when carrying out God's punishment.

Waking Eve, who was disturbed by an evil dream, Adam, "then with voice mild, as when Zephyrus on Flora breathes, her hand soft touching, whispered thus: Awake my fairest, my espous'd, my latest found" (note this reference to ancient mythology in this great Christian epic). A little later, Adam and Eve sing a hymn to the Creator and invite the earth to celebrate his praises: "His praise, ye winds, that from four quarters blow, breath soft or loud; and wave your tops, ye pines." In heaven, surrounding God, are "celestial tabernacles, where [the angels] sleep, fanned by cool winds."[8] On the morning of the fateful day, Adam and Eve "partake the season, prime for sweetest scents and airs."[9]

The Edenic quality of the winds of Paradise change quickly after the Fall. The earth is then devastated by them, which sustain and demonstrate God's anger.

> . . . Now from the north
> Of Norumbega, and Samoed shore,
> Busting their brazen dungeon, armed with ice,
> And snow, and hail, and stormy gust, and thaw,
> Boreas, and Caccias, and Argestes and loud
> Thrascias rend the woods and seas upturn;
> With adverse blast upturns them from the south
> Notus and Afer black, with thunderous clouds
> From Serraliona. Thwart of these as fierce
> Forth rush the Levant and the Povent winds,
> Eurus and Zephyrus, with their lateral noise,
> Sirocco and Libeccho. Thus began
> Outrage from lifeless things . . .

Beast now with beast 'gan war, and fowl with fowl,
And fish with fish . . . [10]

And then all living animals "devoured each other, nor stood much in awe of man, but fled him."[11] This set of attitudes defines *Paradise Lost* and the general triumph of evil. It is inaugurated by the violence of the winds and the evil that they carry. The "sideral blasts" described by Milton are accompanied by "vapor and mist" and, above all, by "exhalation hot, corrupt, and pestilent."[12]

This vivid passage places the detailed description of the winds that blow from all directions at the heart of divine punishment and the triumph of evil, according to an exact geography of the earth – quite surprising since geography was, up until then, the domain of Paradise. The winds were the agents of the earth's devastation and the consequence of the Fall. Their violence is here depicted as the prefiguration of the generalized cruelty that will quickly be established on earth. Milton's genius is to highlight and to bring into play the war of animal against animal.

The strength of Klopstock's *Der Massias* [The Messiah], written between 1748 and 1777,[13] seems equal in the literary domain to the passions of Johann Sebastian Bach (1685–1750) in the musical domain, and the works are nearly contemporary.[14] The epic poem was a major success, especially in Germany. It was still used in the teaching of literature in the middle of the twentieth century, but today it has been so completely forgotten that the better booksellers are unaware of it. However, of all the epics that I have discussed, it is, in my opinion, the one with the greatest impact. Admirable from beginning to end, it relates the Passion of Christ accompanied by the actions and reactions that it provoked in heaven and hell. The strength of its narrative is the result of the presence of celestial and infernal cohorts which frame the events playing out in Jerusalem.

Of course, God surveys, from his retreat in heaven, the great scene that is unfolding on earth. When needed, he unleashes against Satan and sends to the borders of hell the revitalizing wind of his punishment. The storm willed by God poisons the gates of hell, despite all the efforts of the Devil, who dreamed of transforming this abyss of fire into a new earthly paradise. "He strives t' arrest the stormy winds, which drive the tempest on, and lead them (soft subdued to gentlest zephyrs) like balmy breeze." Belial, who dreams of giving these "dismal fields" an appearance "like those fair plains," and "vainly, forever vainly, does he strive the plains, by God's curse wither'd, fresh to deck like Light's fair universe."[15] It was this brutal confrontation between heaven and hell that gave Klopstock's work such an astonishing epic power.

We have already noticed that, in these very different epics inspired by the Bible and antiquity, the winds, as fearsome as they are, are instruments in God's hands or the hands of the gods. They are the ones who release these often devastating forces. We have encountered them as free agents of expression or combat. And it is in this position that they are placed by Jean-Baptiste Grainville (1746–1805), a writer faithful to the spirit of the Enlightenment, in his epic entitled *Le Dernier Homme* [The last man].[16] As the title indicates, it presents an imagery of the wind in a meditation on the end of the world.

Grainville did not have much influence in his lifetime, even though Mary Shelley (1797–1851) noted the originality of his work. In order to understand better the importance of his work in the wind's imagery, let us examine his description of an episode in the war of the winds, which had at last become independent, without being unleashed by any gods:

> Yesterday, there was raised up on our shores a tempest so violent that the terror it caused still endures. I believe that all the winds were unleashed, in a war, one against the other. They have chosen our sky for their battlefield. They rush here and there, without warning, at all points on the horizon. The first shock was so impetuous that it knocked down trees whose roots stretched into hell, and it shook the mountain that sat on the foundations of the earth. As soon as the north winds (aquilons) pushed back the south winds (autans), roaring with rage, the south winds came back against the north winds, stirring them up like waves and taking control of the air. Sometimes all the winds fought at once, causing shock, overturning, uplifting, escaping into whirlwinds, holding steady at the top of hills, hovering over the valleys, and letting go with horrible hisses.[17]

When this tempest abated, the birds appeared. Here again, as always, we find the idea that, in normal times, the winds are enclosed and enchained.

Tasso's *Jerusalem Delivered* is, for its part, a Christian epic about the First Crusade,[18] but the imprint of antiquity is also present, notably – and this might be surprising – in the four winds of Greek mythology: Boreas, Zephyrus, Notus, and Eurus. They are the ones that structure the geographical points in the course of the text, and they have kept the traits that have been attributed to them throughout antiquity. A few examples will suffice to demonstrate this point.

Boreas is designated four times as the north wind in the French translation, and his actions reveal himself decisively: "And brave immoveable, the thousand punches of Boreas rushing on his wintry wings."[19] Furthermore, Jerusalem, defended by Aladin, "crafty in cruelty," is impregnable, "strong on three sides," but "northward intervenes a rampart less secure." "And from his lofty towers," Aladin defies "the coming storm," a metaphor for the north wind and the Christian army.[20] The wind plays a decisive role in the attack led by Godfrey of Bouillon, who is standing by:

When on the sudden rose a friendly blast,
And fierce wild-fire back upon its authors cast.
The winds fought with the flames, and backward blew
The fires; for where the foe the sheds had rear'd
Upon the soft materials swift it flew
Which kindled, cracked, blazed, and disappear'd.[21]

A hymn in praise of Godfrey of Bouillon follows:

... Heaven itself was found
Ranged on thy side; the very winds revered
Thy will, and, summoned by thy trumpet's sound,
Obedient rush'd to war ... [22]

This is an explicit way of praising the role of the winds conceived as a decisive instrument in God's hands as emphasized in the Bible.

However, Boreas was not alone. Together with Notus, "the strong South Wind," "they cuff, they rave, they clash."[23] A little before, Zephyrus, the west wind, had facilitated Godfrey's pious project. While "Heaven seems a sable furnace" and "the frolicsome sweet Zephyr, silent, waving not a wing, his grotto keeps," then, "if winds intrude, 'tis only such as come from hot sands ... blown in stifling gusts."[24] As a result, despair reigned among the crusaders, but Godfrey prayed to the Eternal Father, who launched a thunderbolt and the most impetuous rain.

In the poem, Aquilon, like Boreas, is also a wind from the north, whereas Africus is a southern wind and Eurus is a hot wind. Tasso borrowed the ancient nomenclature for these winds of the earth. The same goes for the sea. Notus, which launched the tempest, finally yielded, "like a cloud before the wind, or azure mist upon the mountain crest."[25]

The influence of ancient literature, especially that of Virgil's *Aeneid*, is clearer in Ronsard's *Franciad* than in *Jerusalem Delivered*. This epic from 1572 celebrates the establishment of the French people by Francus, Hector's son, who came from Troy, just like the Trojan Aeneid founded Rome. From the outset, in this poem, we recognize Homer's influence as well as that of the *chansons de geste* and the poetry of Ludovico Ariosto (1474–1533), the master of Italian literature at the time of the Renaissance.

It would be tedious to cite the very long passages devoted to storms that are scattered throughout the narrative and in the course of which the winds play such an essential role. Let us therefore content ourselves with showing Virgil's influence and the winds' role in the ancient poetic mode of this epic. In effect, Francus' journey is an odyssey punctuated by a series of storms that are the results of the anger of the gods of *The Aeneid*. In his poem, Virgil has Notus, Africus, and Aquilon intervene successively, after Neptune has punctured a hole in the mountainside in order to free the winds. The scenario in *The Franciad* is not much different. Neptune expresses his rancor against Ilion. He intends – which is logical – to take his vengeance through the means of the sea. He mounts his chariot, surrounded by nymphs, and calls forth the winds, the essential actors in the long second book of this epic. At first, Neptune apologizes to them:

> Winds, terror of heaven and the sea,
> I am not the one who shut you inside
> These rocks where, tormented by fear,
> And under a king's sway, you languish in prison.
> Jupiter did it! And without my consent.
> I was unable to resist his power,
> For he is a god of invincible might.[26]

Neptune wants Aeolus to free the winds, "all four of you together," in order to respect "the oath he made me with his right hand."[27]

> May his scepter slice open obscure caverns
> So the winds there enclosed may escape.
> Tell him to release them. And with great noise,
> Charged full of lightning, storm, and night's darkness,
> May they make the sea heave with rage,
> May they vanquish the Trojans with this tempest.[28]

The terrible tempest starts up and leads Francus' Trojans into catastrophe.

What Ronsard described with such forceful detail is therefore a work about the winds, especially the four winds of antiquity. On behalf of Jupiter, Aeolus makes an appeal against their imprisonment, and their anger – here the anger of Neptune – is revealed in terror. They are going to sow devastation and unhappiness. The sea, in its rage, is content to obey, and it is the winds that play the game. Their noisy power is the master of the elements.

> Once the winds had spread over the seas,
> Wave by wave, they caused it to froth,
> Whipping it up from depth to crest,
> An unwelcome and blustering storm,
> Whistling, rumbling, grumbling, and rising
> In great knolls, under the winds' breath,
> Churn after churn, wave above wave,
> Cleaving open the sea to reveal a deep abyss.
> In surge after surge, the sea is raised towards heaven,
> With clash after clash, it is forced down to hell.[29]

All this takes place in the obscurity of "a dreadful night," which robs the sailors of a view of the sea, while a series of lightning flashes bursts through the clouds. Francus prays to Jupiter, but "a proud gust," "a great gale," "a strong blast," all destroy his ships, blowing them about, breaking them down, hitting them from all sides and smashing them to pieces. At the end of this storm, which lasts for three days, the Trojans run aground on "that unknown coast of Provence." Thus, the winds' role was essential in the case of *The Franciad*. They swirled around and, having done their damage, plowed "two churning swirls" into the waters; they "assaulted" the ships and "chased them far."[30]

The winds' role is also essential in Camões' *Lusiads*, the founding epic that celebrates the grandeur of the Portuguese, the people who rounded the Cape of Good

Hope and reached India after numerous attempts.[31] At the same time, though, they carried the Word of God, an objective proclaimed by Camões but not well developed in the text. In the course of this journey, as in all others from modern times, actions are determined by the winds, but there is more.

Commentators have taken note in this immense epic of the severity of the skies, the winds, and the storms, all allied directly with one another. The Portuguese know how "to brave the wrath of Aeolus' children" and how to open up unknown routes. In brief, there is no doubt: those who are opposed to the sailors and their designs are the winds associated in the course of several events with deceitful and cruel men, the infidels. Despite the Catholic faith, as proclaimed by the author, it is the ancient gods and the mythological winds that play a leading role.

The enemy of the founding hero Lusus and his people is Bacchus, seated in Olympus. He wishes to be the one and only conqueror of India. In a word, he is the one that the Portuguese intend to dethrone. As for Venus, she defends them. The quarrel of these two gods disrupts Olympus. Thus, "when the winds rage fiercely through the dark forest, leaving a trail of devastation in their wake, the whole countryside clamors and re-echoes with the commotion."[32] We see here that Camões, in evoking the combat in Olympus, does not neglect to describe the earthly tempest let loose by the winds. Nor does he neglect to indicate the tenor of the winds. When they are calm, he writes, "in their dark lairs, the raging winds lay at rest."[33] This shows the strength of the winds' imagery as enchained and enclosed in a rocky cave or inside a wineskin, as we have read in *The*

Odyssey. The naval battles – such as the one against the Moors in the first canto – are actually fought against both Bacchus and Venus.

Lusus, spreading "our sails to the wind,"[34] is struck by a typhoon. Let us remember that Camões was writing in 1559 and that the phenomenon, although described by Pliny the Elder, would be further described by James Cook a little later on. The reader of the twenty-first century should not be surprised by the description of this meteorological phenomenon, since he is saturated with televised images. That is why I cannot resist reproducing Camões' very long description – a description given at a time when this atmospheric phenomenon, this "very stuff of clouds," then totally unexplained, must have been so little known.

> One – I had a clear view of it – was our blessed St. Elmo's Fire, that is sometimes to be seen in times of storm and raging winds, when the heavens lower and even strong men give way to tears. And it was no less demonstrably a miracle to all of us, a thing to strike terror to our hearts, to see the clouds drinking up, as through a long spout, the waters of the ocean.
>
> First, a thin smoky vapor formed in the air and began to swirl in the breeze; then out of it there took shape a kind of tube stretching right up to the sky, but so slender that one had to strain one's eye to see it – made, as it were, of the very stuff of clouds. Gradually it grew and swelled until it was thicker than a masthead, bulging here, narrowing there, as it sucked up the water in mighty gulps, and swaying with the ocean swell. At its summit, a thick cloud formed, that weighed heavier and heavier with the mass of water it absorbed.

Sometimes a beast, drinking rashly from an inviting spring, will pick up a leech that fastens on its lips and there sates its thirst with the animal's blood: it sucks and sucks, and swells and swells. In the same way, this mighty column waxed ever mightier, and with it the black cloud that crowned it; until at length, sated too, it drew up its lower extremity from the sea and drifted off across the sky, sputtering the ocean with its own water returned as rain, restoring to the waves what it had stolen, minus its salty savor. And now let the experts consult their authorities and explain to me, if they can, these mysteries of nature.

Had the philosophers of old, who journeyed through so many lands to learn their secrets, witnessed the wonders that I have witnessed as I sailed hither and thither over the waters, what writings would they not have left us, what revelations concerning the workings of the stars in their courses and the many marvels and properties of nature, and every word the naked, unvarnished truth.[35]

As to the storms' spirit, this "mighty column," which holds sway over the headlands of the cape, Camões identifies him with the last of the giants conquered by the mythological gods. In *The Lusiads*, it is he who lives at the end of the African continent and sees as far as the Antarctic pole, boasting of his control over the winds.

The journey of the Portuguese, after rounding the cape, continued on toward India, even though they were opposed by Notus in his anger but aided by "the sweet breath of Zephyrus."[36] Camões, returning to ancient mythology, interrupts himself here by describing in detail in his sixth canto the role that Neptune, in his palace surrounded by his courtiers, played in the actions of the four winds – and this is what concerns us here.

In order to antagonize the Portuguese, this god sends a message to Aeolus, "bidding him to loosen the countless furies of the conflicting winds and sweep the seas clear of mariners once and for all." There then follows a page devoted to their actions – something that confirms their constant presence and importance in this epic from the Renaissance. Meanwhile, before the winds were unleashed, the fleet of Lusus' children was "being gently wafted onward on its long voyage."[37]

Bacchus, however, is not discouraged, and this time he raises up and unleashes the winds. "The winds raged and roared, whistling through the shrouds and lashing the storm into ever greater fury." Venus, who notices this, sends her nymphs, and "no sooner had they come in sight than the vigor with which the winds had till then been waging combat forsook them, and they tendered obedience as though beaten in the fight."[38] Then,

> The most fair Orithyia addressed herself to Boreas, the one she was fondest of at heart . . . Beauteous Galatea spoke in the same strain to Notus, knowing well that he had long delighted in the sight of her and confident that she could do with him what she would . . . The other nymphs wrought a similar sudden change in their respective adorers, and soon, all wrath and fury spent, the surrender to lovely Venus was complete.[39]

In conclusion, the certainty that Venus will preserve the loves of Aeolus' children and that they will respect the Portuguese, the goddess's "favorites," is essential for the rest of their travels.

These incessant references to ancient mythology – to Bacchus, Venus, Neptune, and Aeolus – in a Christian epic does not fail to astonish. However, it confirms the importance of our subject: the destiny and the action of the winds, each of them with their own personality. The explanation of this constant presence, which might appear strange, is connected with the respect for *The Aeneid* in the sixteenth century – an admiration avowed by the readers of Camões as well as the author himself.

Upon arriving in India, specifically at Calcutta, "where her course lay, the wind was calmed, and only the graceful motion of her passage disturbed the air."[40] The state of the atmosphere symbolizes the journey's success, and this victory is prolonged under "the sweet breath of Zephyrus," which animates the flowers and welcomes the nymphs who are directed toward the palace of Tethys (the sea). Vasco da Gama, the leader of the expedition, hurries to join them.

In the eighteenth century, Jean-François de La Harpe (1739–1803) was very happy to censure the presence of pagan divinities in this epic, because, according to him, its subject was, first and foremost, the triumph and establishment of Christianity.[41] We now understand that position very well.

7

The Fantasy of the Wind in the Enlightenment

At the end of the eighteenth century, two works, which are not properly speaking epics but which could still qualify, exercised a great deal of influence on the representation of nature: these are the poems of Ossian, purportedly discovered by the Scottish writer James Macpherson (1736–1796) and first published in 1769, followed by other poems in 1773, and *The Seasons* by James Thomson (1700–1748), first published in 1730 but popularized in France in 1769 by a poem of the same title by Jean-François de Saint-Lambert (1716–1803).

The poems of Ossian are a series of verses attributed to a bard whose work James Macpherson claimed, probably falsely, to have collected and translated. They were inspired by an ancient literature emanating, presumably, from Ireland and Scotland. This work, which conveyed the root evils that tormented that time, recast the wind's imagery in its own fashion. German Romanticism was strongly "ossianated." Herder transmitted to Goethe his admiration for the works of this "Caledonian" bard of the third century; Schiller admired them; and Klopstock, we are told, was reading them on his deathbed. Almost all the great German composers – Schubert, Mendelssohn, and Brahms – portrayed Ossian in their music, and his spirit was appropriated by Casper David Friedrich in his paintings. The authors of English Gothic novels owe him a great deal. Blake, Coleridge, and Byron – and later on the Brontë sisters – have all come under his influence.

And what can we say about France? Artists of the empire – and the emperor himself – had great admiration for Ossian. They compared him to Homer. Enthusiasts today remember the Ossianic heroes of Girodet. According to Yvon Le Scanff, in the works of the great writers of the end of the eighteenth and the beginning of the nineteenth century, the *locus horridus*, the frightful place, as opposed to the *locus amoenus*, the idyllic place, was a frequent motif of a chaotic landscape where the winds blow.[1] Very early on, Denis Diderot (1713–1784) became the translator of these fragments cited – or created – by Macpherson, the true author of these poems. The somber energy of the wind was further propagated in subsequent Gothic works. Mme Germaine de Staël (1766–1817) established a parallel between Homer and Ossian, whose sensibility was pleasing to "the wind's noise in their savage heaths." She recalled that the "Caledonian" landscape remained for a long time the archetype of the sublime landscape.

Alphonse de Lamartine (1790–1869) evoked Ossian as "the poet of waves, of the inarticulate plaintive cry of the North Sea."[2] Chateaubriand, who cited Ossian in six of his works, wrote the following famous passage in his hymn to the Gothic style, *Genie de Christianisme* [Genius of Christianity]:

Under a cloudy sky, amid wind and storm, on the coast of that sea whose tempests were sung by Ossian ... , seated on a shattered altar in the Orkneys, the traveler is astonished at the dreariness of these places ... The wind circulates among the ruins, and innumerable crevices are so many pipes which leave a thousand sighs.[3]

Astolphe de Custine (1790–1857) and Étienne Pivert de Senancour (1770–1846) reinforced these aphorisms: "The Ossianic landscape illustrates the sickness of the age," writes Le Scanff.[4] "The melancholy of Ossianism" and its "sad despondency,"[5] all born out of the contemplation of nature, were cited by Alfred de Vigny (1797–1863) in a passage from Macpherson's *Fragments of Ancient Poetry*: "Rise, winds, rise; blow upon the dark heath."[6]

Ossianism seems "like a barbarous regeneration," writes Le Scanff. "The sublime of Ossainism is conceived as the expression of primitive nature and as the oblivion of art and cultural artifice that weakens the enthusiasm, strength, and simplicity of origins." The incoherence and disorder of nature, the sublime character of the storm, "the terrible capacity for physical and mental debilitation" are all linked to Ossianism.[7] In the opinion of our contemporaries, the Ossianic tempest is, first and foremost, a maritime storm. The presence of the wind in Macpherson's poems is, in the beginning, a puff of wind, more often than not coming from the north, from "the mountain of the winds" or "the hill of the winds." It crosses the moor, assailing the heather and the flowers, and then goes off toward the rocks on the edge of the sea. There the dead heroes lie and are mourned by young girls, who were certainly promised to them. In fact, in this place, the wind encounters the fiancée bathed in tears, and it makes a lugubrious sound that mixes with the cries and the tears of the woman.

In Ossian's poems, blowing almost always in a somber environment – that is to say, at night – the wind is linked to death – to the death of the heroes from long ago, such as Fingal or Ossian himself. Their ghosts continue to haunt the memory of old men, and so do the ghosts of heroes recently deceased whose deaths arouse the tears – and sometimes the suicides – of young women. The Ossianic movement is all about physical brutality, death being present without exception. The plaintive cry – an essential element in this poetry – is inseparable from action. Heroes roar, fight each other savagely, and fall like battered oaks. Heroines are like heavenly creatures.

Let us return to the winds in more detail. On several occasions, the author, like the heroes and heroines, urges and invites the winds to chase away the clouds that hide the countryside, the mountains, and the sea. However, it so happens in *The Songs of Selma*, one of the cycles of poems in Ossian's works, that the stormy winds calm themselves in order to let the bright evening star appear. Sometimes the wind is called forth by the cries and tears of women who are prostrated on the tombs of the heroes. As we have already said, in autumn and in winter the wind is described as ravaging the moor. It whistles in the heather and the grass that it assails. Sometimes the vegetation itself cries out.

The wind appears, above all else, as a voice, often loud, and man – or more often woman – appeals to it to transmit messages. In another role, the wind revives the memory of dead heroes. It incites reminiscences and fidelity to memory. Zephyrus, more rarely evoked than Boreas, is there to celebrate the voice and the virginal beauty of women. Let us cite a few short fragments:

> Evening is grey on the hills. The north wind resounds through the woods. White clouds rise in the sky. The trembling snow descends ... Sad, by a hollow rock, the grey-hair'd Carryl sat ... Clear to the roaring winds he lifts his voice of woe ... He, the hope of the isles, Malcolm, the support of the poor, foe to the proud in arms ... It is he! It is the ghost of Malcolm! Rest, lovely soul, rest on the rock ... Why, ye winds, did you bear him on the desert rock?[8]

In Fragment VIII, Ossian exclaims: "Such, Fingal, were thy words, but thy words I hear no more ... I hear the wind in the wood, but no more I hear my friends."[9]

A young woman wants to hear her lover Schalgar's voice and declares: "It is night, and I am alone, forlorn on the hill of storms. The wind is heard in the mountain. The forest shrieks down the rock." She discovers Schalgar and his brothers dead, and she addresses them: "Oh! From the rock of the hill, from the top of the mountain of winds, speak, ye ghosts of the dead, speak ... Whither are ye gone to rest? In what cave of the hill shall I find you?" But the wind does not bring her an answer. "When night comes on the hill, when the wind is up on the heath, my ghost shall stand in the wind and mourn the death of my friends."[10]

In Fragment XII, we read Ryno's question regarding Alpin's melancholy: "Why complainest thou, as a blast in the wind?" Alpin responds: "A tree with scarce a leaf, long grass which whistles in the wind, mark to the hunter's eye the grave of the mighty Morar."[11]

In Fragment XIV, Duchommar expresses his love of Morna: "But thou art like snow on the heath, thy hair like a thin cloud of gold on the top of Cromleach, thy breasts like two smooth rocks on the hill ... " But she does not love him; she loves Cadmor, whom Duchommar has slain: He declares: "High on the hill I will raise his tomb, daughter of Cormac-Carbre. But love thou the son of Mugruch," he commands, "his arm is strong as a storm."[12] Here is an allusion to the participation of the winds in the argument, which is a *leitmotiv* of Ossianic poetry.

Most of the commentators on Thomson's *Seasons* have not gone beyond the last part, "Winter." They are wrong to do so, as we will see in the next chapter. Having said this, let us concentrate for the moment on this season, the evocation of which has often been praised and to which is attributed a great influence on the literature of landscape. According to Thomson, the fearsome wind, which is the companion of winter, is the north wind, the wind of rage, and, when he begins his description of this last season of the year, he begins by addressing the winds, which he expects to launch a tempest. At the same time, he asks himself: Where is the retreat from which the winds surge forth?

> Ye too ye winds! that now begin to blow
> With boisterous sweep, I raise my voice to you.
> Where are your shores, ye powerful beings! say,
> Where are aerial magazines reserv'd
> To swell the brooding terrors of the storm?
> In what far-distant region of the sky,
> Hush'd in deep silence, sleep ye when 'tis calm?[13]

The words are clear: The tempest – as exalted by Thomson and repeated by his flatterers – is obedient only to the winds. Its origin is their unknown retreat. It is formed, more often than not, at night, as if pushed by a nocturnal demon. It is then discovered by its cries and sighs. Then it is unleashed. Thomson launches into an all-too-brief description:

> Huge uproar lords it wide. The clouds commix'd
> With stars swift gliding sweep along the sky.
> All Nature reels . . . [14]

. . . until God, "Nature's King," commands it to be calm.

Nevertheless, there is another natural disaster that is due to the wind, and it is not the tempest:

> With the fierce rage of winter deep suffus'd,
> An icy gale, oft shifting, o'er the pool
> Breathes a blue film, and in its mid career
> Arrests the bickering stream . . . [15]

Here is an allusion to the immobility of frost. And so Thomson evokes the Laplanders, who live in a place where

> . . . Winter holds his unrejoicing court
> And thru' his airy hall the loud misrule
> Of driving tempest is forever heard.
> Here the grim tyrant mediates his wrath;
> Here arms his winds with all subduing frost,
> Molds his fierce hail, and treasures up his snows
> With which he now oppresses half the globe.[16]

Perhaps the most interesting point is that winter, like all the other seasons, responds only to the attributes, feelings, and designs of the Creator. Thomson's work, which belongs to descriptive poetry that exalts the feelings of nature, ends with a hymn to the Divine Being:

> These, as they change, Almighty Father, these
> Are but the *varied* God. The rolling year
> Is full of Thee. Forth in the pleasing Spring
> Thy beauty walks . . .
> Then comes Thy glory in the Summer-months,
> With light and heat refulgent . . . [17]

In autumn, it is Zephyrus who is the face of the Divine. Then there follows a scene that we have encountered several times since antiquity:

> . . . On the whirlwind's wing,
> Riding sublime, Thou bidd'st the world adore,
> And humblest Nature with Thy northern blast.[18]

Thomson details the roles of certain winds in this hymn, which all reveal the faces of God:

> . . . To Him, ye vocal gales,
> Breathe soft, whose Spirit in your freshness breathes:
> Oh talk of Him in solitary glooms!
> Where, o'er the rock, the scarcely waving pine
> Fills the brown shade with a religious awe.[19]

And the author at last addresses himself to the fiery southern winds:

> And ye, whose bolder note is heard afar,
> Who shake th' astonish'd world, lift high to heaven
> Th' impetuous song, and say from whom you rage.[20]

The text is clear, and the interpretation is easy: the seasons are the different faces of God. And it is the winds – not the sea – that obey him and express his tranquillity as much as his wrath. At times, they even express the ethereal ambiance that reigns in the earthly paradise. It is this latter point that we must consider in the guise of the interlude devoted to Zephyrus.

8

Gentle Breezes and Caressing Currents

History has scarcely remembered anything but "Winter" from Thomson's *Seasons*, but, for what concerns us, "Spring" and "Autumn" are also very interesting and, in a way, much more important. In effect, he highlights in these two seasons the importance of gentle breezes and caressing currents, both of which are in harmony with the rhythms of vegetation and the pleasure of nature. In this, Thomson revives the genre of the ancient idyll, that of Theocritus, the Greek poet of the Hellenistic era, whose bucolic work has had an enormous influence.

The west wind, as we have read in the epics, could be – but rarely – menacing. Its image is contradictory. It remains, above all, however, the wind that blows lightly and refreshes sweetly, the wind of subtle pleasure in all its forms, the revealing wind of feminine thrills, and the wind that favors amorous sensuality and the meetings of lovers. All of this has already been emphasized in baroque poetry, as demonstrated by Véronique Adam. According to her, in this poetry, the wind found its voice and made itself the double of the lover out of necessity.

It could brush against the beloved without her rebelling, and it could facilitate erotic dreams. Moreover, by its unpredictability, it embodied feminine inconstancy because it was flighty by nature.[1]

Observing that the woman's body cannot resist, the wind could enflame her lover. For evidence, she allows herself to be touched; it caresses her eyes, her hair, and her mouth. In a word, everything seduces the one who is in love. The west wind, especially in spring, is the wind for lovers. It seems sometimes to express sympathy for those in love who are suffering, and it receives their plaintive cries. Claude de Trellon (d. 1610 or 1625) wrote in 1595:

> Wind, may you be content to kiss whenever you like
> The mouth and the eyes of my beautiful warrior princess!
> Why am I not like the wind so that my imprisoned soul
> Could presently break its shackles, its bonds, and its vows![2]

And Étienne Durand (1585–1618) wrote a few years later: "I would like to be the wind sometimes / To play in the hair of Urania . . . "[3]

Painting, like literature, celebrates the loves of Zephyrus and Flora. It is this wind, the west wind, that seems to be the voice that peddles at the same time the odor of nature and the beauty of woman. This is what Sandro Botticelli (1444–1510) expresses so well in his famous painting *Springtime*. In a word, the west wind possesses a tactile, olfactive, and visual existence.

Régine Detambel, in her *Petit éloge de la peau* [Little elegy to skin] analyzes in detail the effects of the wind on the epidermis. She writes that the wind tickles "this little bit of skin that drives all senses of a person crazy." She brings to light the multiple forms of those "light touches" and "rustlings," and she reminds us that women enjoy touch much more than sight. "Under the act of caressing, the skin breathes; it palpitates; it enters and leaves its own skin without ceasing." Furthermore, "in the act of caressing, the skin does not resist or oppose anything." "In the act of caressing," she concludes, "the ephemeral does not end." What brings Detambel to our attention is the following: "It was necessary to invent an imagery of the wind in which the act of caressing would be wind color, wind wisdom, or wind echo – an invisible thing without opacity at the limits of sensibility."[4]

Let us come back to the west wind of spring and autumn in Thomson's poem. In *The Seasons*, it is a witness to harmony and the absence of corruption in the air: "Clear shone the skies, cool'd with eternal gales, / And balmy spirits all . . . "[5] In their own way, the "western breezes" re-create the earthly paradise, in which

Pure was the temperate air; an even calm
Perpetual reign'd, save what the Zephyrs bland
Breath'd o'er the blue expanse . . .[6]

And Thomson was sensitive to subtleties. He wished that his song

. . . may perfume my lays
With that fine oil, those aromatic gales,
That inexhaustive flow continual round.[7]

He exalted these perfumes that "A fuller gale of joy, . . . liberal, thence / Breathes thro' the sense, and takes the ravish'd soul."[8] The west wind differed from the gale, which, in its brutality, ignored all subtlety, all delicateness, and all enchantment.

Elsewhere, Thomson intones a hymn to the breezes of May:

A potent gale, delicious, as the breath
Of *Maia* to the love-sick shepherdess
On violets diffus'd, while soft she hears
Her panting shepherd stealing to her arms.[9]

And then the lover arrives . . . Here the author does not escape the stereotype that we have seen in the writings of the baroque poets.

At the awakening of summer, "the Zephyrs floating loose" could make themselves felt and several "ever-fanning breezes" would blow sometimes during "the sultry hours." In this season appeared a woman bathing, a "fairer nymph," and the sensuality of her lover, her "swain," was "thrice happy" when he surprised her. "And robed in loose array, she came to bathe her fervent limbs in the refreshing stream." She "stripp'd her beauteous limbs to taste the coolness of the flood," but, "with fancy blushing at the doubtful breeze alarm'd," she hesitated.[10] In Autumn, the gentle breezes and the western winds return.

At the end of the eighteenth century, Thomson was not the only poet who exalted the western winds. There was also, as we have already noted, the winds praised by Ossian, as well as the work of Salomon Gessner (1730–1788), who tirelessly took up this theme, notably in his *Neuen Idyllen* [New idylls].[11]

In one poem entitled "The Zephyrs," the first breeze invites the second one to flutter about near the nymphs, but the latter refuses: "I shall imbrue my wings in the dew that bathes these flowers and gather their delicious perfumes." This is intended for Melinda, a beautiful girl, who is soon walking along the footpath. In a word, then, this breeze is in love: "As soon as she appears, I fly to meet her; my wings, spreading round her the most sweet perfumes, will cool her burning cheeks; while I kiss the tears just starting from her eyes." At these words, the first breeze decides to imitate his companion. When Melinda appears, he declares: "I will, like thee, imbrue my wings in the dew that bathes these flowers; like thee, I will gather their perfumes; and like thee, at the return of Melinda, I will fly to meet her."[12] Such is an example of the personification of the winds in love, and this dialogue between the two breezes constitutes the idyll.

The winds in desiring women are evoked several times in Gessner's poetry. In the idyll entitled "Thyrsis," he writes that the breezes in their games try hard to uncover Chloe's young breast: "her airy garment, winding in graceful folds about her shape and knees, behind her floated, at the pleasure of the winds, with pleasing murmurs."[13] In our century, in scenes where tempests blow, it is violent winds that cause clothing, not to flutter, but to plague women's bodies with ruin.

The return of the idyll in the middle of the nineteenth century in the *Poèmes antiques* of Charles-Marie Leconte de Lisle (1818–1894) is well known. We will emphasize here only one essential point, since we are going to examine his poetry in more detail in the following chapter. In his "Aeolides," a hymn to the virgin daughters of Aeolus and to the delightful sensuality of the breezes, he writes:

Oh, breezes, floating in the air,
Sweet breaths of beautiful springtime,

Who dole out capricious kisses.

. . .

Caress the mountains and the plains!
And the air where your flight murmurs,
Filled with aroma and harmony.[14]

This poem inspired the composer César Franck (1822–1890).

As to other breezes, "in flight so fresh," according to Leconte de Lisle, they are like "immortal laughs which fill the earth."

Oh! How many arms or adored shoulders
Have you already kissed,
At the edge of the sacred fountains,
On the hill with the forested side?[15]

And the poem ends with a prayer addressed to the ancient Aeolides: "Breezes of the lesser gods, visit us still . . . "[16]

Nevertheless, the most interesting point resides in the outline of another poem:

Oh, breezes! Who came from the heavens!
And laugh at all things!
With your capricious kisses
Why do you ravish the scent of roses?

. . .

You are the breezes of the heart,
Full of illusions, kisses, and sighs,
And when our souls are full,
You flee, singing sweetly and jocularly![17]

Edgard Pich thinks that this poem reveals a nostalgia for a state of youthfulness in Leconte de Lisle, who tends to present antiquity as a place of happiness.[18]

The gentle breeze, the zephyr, lifting up the woman's dress, revealing her body, and making her both desired and desiring, was used in an entirely different context by Gustave Flaubert (1821–1880). It is this flowing current made by the wind that made St Anthony's temptresses particularly desiring and desirable when they appeared to him: "The wind passing through the intervals between the rocks, makes modulations; and in those confused sonorities he distinguished Voices, as though the air itself was speaking. They are low, insinuating, and hissing."[19]

As an exception to these texts from the eighteenth and nineteenth centuries, and in order not to set aside the erotic power of the wind, let us now turn to a twentieth-century text written by Jean Giono (1895–1970) in his novel *Regain* [Harvest], because there he describes forcefully the sensuality of the wind as it works on a woman's body. The scene unfolds on "the plateau," when the winds are strong and blow without respite on the heather. This time, it is not a gentle breeze but a gust of wind that penetrates Arsule's body – the woman coming to this place with her lover Gédémus. It does so as a total sexual union.

> As soon as they had got up and stepped along the track, they had to reckon with the wind. It was blowing full in their faces and clapped its big warm hand on their mouths, as if to prevent them from breathing. They were used to it. They just turned their faces round a little, to drink in the air on one side, as swimmers do, and thus moved on a good way. It was tiring, but not so bad. The wind began to scratch their eyes with its nails. Then it tried to tear off their clothes; it nearly blew off Gédémus' coat. Arsule was

pulling on the strap, leaning forward. The wind entered her bodice, as if at home there. It flowed between her breasts and stole down to her belly as might a hand; it flowed between her loins, cooling her like a bath. Her back and hips were all freshened with the wind. She felt its freshness on her, but also its warmth, as if it were full of flowers, and its tickling, as if she were being stripped with handfuls of hay. That is what people do during the hay-making season. It puts some women all in a flutter, as the men well know. Then, suddenly, she began to think about men. It was the wind that had played the part of a man for a while.[20]

Gédémus leaped toward Arsule, stimulated by the caresses of the wind: "He seemed to be disturbed. Arsule looked round at him with a tender and caressing eye ... Her body was fermenting like new wine." Arriving at Trinity, "a hamlet packed in the midst of the plateau," she was tired; "there was no longer a wind to caress her," only "a hush all round." Despite this, she still thinks about the man: "It seemed to her that the fingers of the wind were still upon her and that its great hand was laid on her bare flesh."[21]

In the following two days, Arsule remains agitated by this man's desire – a desire that was brought to life by the wind on the plain. Consequently, she cannot sleep because of this "need." Later, she gives herself, not to Gédémus, but to the giant Panturle, who satisfies her.[22]

In no other text, to my knowledge, is the wind so closely associated with the appearance of aggressive desire. The wind has a man's knowledge of how to excite a woman and lead her to sexual abandon. It begins with foreplay, according to ancient peasant sexual practices, and it ends with the total possession of the woman's body, almost an explicit identification of the wind as a rapist. Because of it, this desire has become irrepressible. It remains in silence. It is necessary to cite this passage because it unites the gust of the wind, more so than the gentle breezes, with the delirious desire of the woman.

9

The Enigma of the Wind in the Nineteenth Century

At the height of the nineteenth century, the references to the wind's imagery are infinite. To continue our Aeolian promenade, we must therefore make a choice. We will concentrate on three authors of the French language whose texts are devoted to the wind and seem to me to have the greatest impact: Victor Hugo, of course; Leconte de Lisle, whom we have already encountered briefly in the previous chapter; and Émile Verhaeren.

The first of these, as everyone knows, was obsessed with winds and storms. In a word, his work marries personal experiences with the play of the imagination. However, the wind is not a theme to which he attached much importance in his early years. Its presence intensified in his writings from the time he went into exile on Guernsey, and it did not weaken until the end of his life. We know that it was on this island fortress that he composed *Travailleurs de la mer* [Toilers of the sea], which he did not publish until after his return to Paris. It was during a storm on the night of February 14,

1856, that Hugo, in a state of semi-consciousness, wrote the following:

> And how the clarion-blowing wind unties
> Above their heads the tresses of the storms:
> Perchance even now the child, the husband, dies.
> For we can never tell where they may be
> Who, to make head against tide and gales,
> Between them and the starless, soulless sea,
> Have but one bit of plank, with one poor sail.[1]

Since it is necessary to recall the experience of the wind as rendered by Hugo, installed as he was on his island in the English Channel, it is also necessary to stress that the essential point rests in the interpretation that he gave: he identified man with the wind.

Nevertheless, it seems that the most appropriate notion for defining the wind in Hugo's mind is that which links it to the nocturnal. One cannot, in his work, study the one without the other. Yvon Le Scanff has put it this way: "The force that disturbs the sea and the night is that of the winds, which seem to come from infinity. They have the power to confuse."[2] Hugo writes: "The winds from the open ocean . . . They hold the dictatorship of chaos. They control chaos." That is to say, chaos had its revenge on creation. The storm's winds succeed in piling up, creating an "oscillation of two oceans, one upon the other" (i.e. the ocean of air superimposed on the ocean of water), and this confusing exchange is chaos. The wind proves that "a remnant of the agony of chaos remains in creation."[3] It manifests itself in "the anger of the unknown," "the roar of the abyss," "the great brutish howl of the universe," and "the inarticulate finding utterance in the indefinite."

The storm with its winds, this "strange cry, prolonged and continuous," constituted "a complaint." "The void bewails and justifies itself."[4]

In *Toilers of the Sea*, Hugo describes the winds in detail: Like children, "the winds rush, fly, swoop down, finish, begin again, soar, hiss, roar, laugh . . . taking their ease." "The frightful thing about them is that they are playing. Theirs is a colossal joy, composed of shadow."[5] At all times, Hugo does not judge; he does not denigrate, because, as Le Scanff writes, "the dynamic nature of the sublime is the same image as the inspired creator."[6] In this way, Hugo demonstrates his opposition to the somber nature of the sublime in Ossianic poetry or the colossal nature of the sublime on the mountain.

The other face of Hugo's imagery of the wind is the enigma, as demonstrated by Françoise Chenet. The wind is, by its very essence, a strange, insistent, and disagreeable voice. It operates through its reiteration and its "implacable instability," which is that of an unknown force, forever impenetrable. The wind is an "element that makes the invisible palpable," but "the extravagance of the air," like all that reveals the unknown, incites dreams.[7] Such is the second tenet of the wind in Hugo's imagination. Above all, let us repeat, it is an enigma. It manifests itself in "the ramblings of the abyss." What does the wind say? To whom does it speak? Who is its interlocutor? Into whose ear does it murmur?[8] As an enigma, the wind irritates Hugo, who entreats it to take on another form: "Why this whistling, always the same? Why this creaking, always the same? What good does it do to shout oneself hoarse in the clouds, always repeating the same things? Please change your exclamations."[9]

Chenet concludes her study with a breath-taking affirmation: in the end, Hugo identifies with the wind and makes it a major part of his personal mythology. To tame the winds in order to go ever further in this ocean in the air – something that confronted the balloonists at this time – was the most audacious enterprise. The "full sky" of *The Legend of the Centuries* exalts the ultimate feat of man, this "traveler of the infinite," who takes off in his "magic and supreme vessel . . . that bounces on the waves of the wind."[10]

Finally, let us remember the words from the shadow's mouth in the last poem of *Les Contemplations*, "Everyone Speaks." Even after a fall in prison behind bars:

The soul makes out from afar the eternal glow:
In the trees, it quivers, and, without day or without sight,
It feels, still in the wind, something from the heavens.[11]

We have already touched, however briefly, on the relationship between Leconte de Lisle and the wind, as it is evoked as the subtle sweetness of the breezes or western winds in his *Ancient Poems*, but this is of secondary importance. The real discussion of the winds by this poet is in his *Poèmes tragiques*, which are too often forgotten. He presents their power, their violence, their persistent rage, and their memory with a talent that is not found anywhere else in his work. And it is not just storms and hurricanes about which he writes. The awful winds here are those on land, moved by a violence that is one of vengeance and punishment, at times the bringer of horror, injustice, and unhappiness.

A few brief fragments will illustrate each of the facets of the power with which this poet identifies the winds' actions. We may recall Baudelaire's albatross, but that poem is lacking in the violence with which Leconte de Lisle presents this bird in its battle with the winds, which seem to be the main character in the poem.

In the great expanse from Capricorn to the Pole
The wind hollers, roars, whistles, moans, and mews,
And it leaps across the Atlantic all white
With furious spittle. It pounces, scratching
The pale waters that it pursues and dissipates in mists;
It bites, tears, lifts up, and cuts off the stormy clouds
Through convulsive stretches where a sudden burst of
    light bleeds forth;
It seizes, envelops, and tumbles in the air.
A whirling mix of bitter cries and plumes
That it shakes and drags on foaming crests,
And hammering on the massive heads of sperm whales,
It mixes its howls with their monstrous sobs.
Alone, the king of this space, of the seas without shores,
Flies against the assault of these savage gusts of wind.
. . .
[The albatross] breaks through the whirlwind of raucous
    expanses,
And, quietly in the midst of this appalling terror,
Comes and goes and disappears majestically.[12]

By contrast, the accumulation of words that transcribe the actions of these "savage gusts of wind," along with their rage and cruelty, has a way of making the animal, "alone, the king of this space, of the seas without shores," more heroic. Its quiet majesty defies "this appalling terror."

Let us come back to the idea of the wind as the bringer of punishment and/or remorse. The longest passage in these *Tragic Poems* – and without a doubt

the most grandiose – is entitled "Magnus' Greyhound."
This man, Magnus, fought as a knight during the
Crusades, but he did not attack the Saracens. Instead,
he committed horrors in the Holy Land, practicing
thievery, as we learn later, on the crusaders themselves.
In the poem, Leconte de Lisle presents him, in company
with his dog, after his return to his castle in the country
of his birth. But the wind, despite the distance that seems
to protect him, has not forgotten any of his misdeeds. It
takes its revenge and it destroys. This text, the longest
in the volume, with its hallucinations, corresponds well
with the title of the work.

> And the wind chases its tomb-like moan forward,
> From the depth of its prison to its shaking summit,
> Rising up on its staircase that twists itself into a spiral.
> With a hoarse howl mixed with bitter cries,
> It fills the open crevices in the wall
> And flogs the doorknob on its loosened hinges.
> It rattles the iron railings to their foundations,
> Or, sometimes, squatting in the dark corners,
> It pushes out a bitter laugh like a demon who scoffs.
> Duke Magnus heard not the cries nor the leaps and
>   bounds
> Of the wind which strove forth across the debris
> And knocked over the owls with rounded eyes.[13]

And, further on, he writes: "The wind always howls out
of doors and rages."[14] It is self-evident that remorse,
which tortures Magnus, will henceforth never cease. It is
stirred up by a wind that howls and rages without end.

There is another horror carried by the wind in
these *Tragic Poems*, one that comes through in "The
Coronation of Paris," in the midst of which the wind

and its gales are the symbol of the unhappy people of the capital and their resistance against the Germans in the siege of 1870–1. At that time, the wind did not cease to howl, to terrorize, and to destroy, but it also made the city and its people into heroes. In some mysterious way, it was the manifestation of the sacred. The somber, nocturnal, furious, and hateful character of the wind that beat and blew and scowled only caused the unshakeable confidence of the Parisians to be magnified in the face of these gales. It is well known that this text was written in January 1871 when Paris was still resisting the Germans:

> The bitter wind, which jumps over the hill and the plain,
> Comes, filled with execrations,
> Supreme fury, vengeance, and hatred,
> To strike the somber bastions.
> It flogs the heavy cannons, the giant packs
> That sit up stretched out along the carriages.
> It blows at times in their gaping mouths
> That it fills with a confusing moan.
> It shakes the rooftops, with snow-covered debris.
> An immense sepulcher, already closed,
> From where murmurs made up of sobs,
> Lamentable and numberless, show up still.[15]

And, a little later, there is a hymn to Paris:

> Oh, unwavering nave in the waves, like gales,
> Which, under the black or mild skies,
> Joyous, and deploying your triumphant sails,
> Voyaging forth victoriously.[16]

After reading this poem, we should recall the role attributed to the wind during the "Bloody Week" of

May 1871. The archives at the Paris Observatory show that it was a wind from the west that blew during these days, and it fanned the flames of the buildings set ablaze by the partisans of the Commune, who then retreated toward the east. The Versailles press did not fail to highlight this meteorological event, which they interpreted as a sign of divine intervention.

Very different is the wind's imagery that comes across in Verhaeren's poem – a poem that is specifically devoted to the "wind." It was written in the Symbolist vein, which was the author's style. He did not intend to describe the fury of the gales, but he made the wind a symbol of the gloom and the sadness of the countryside in northern France. He wrote a series of poems entitled *Les Villages illusoires* [The illusory villages] in sympathy with the sad state of nature with which humankind is identified. This collection of poems, according to Warren Lambersy, evokes the north wind as one that blows "in a country where there is nothing to see," a space "without natural boundaries," an "immaterial prison of the infinite." These poems, especially the one entitled "The Wind," are symbols of a "world dedicated to suffering and unhappiness." The poet feels himself "integrated into this suffering world" that reveals to him the essence of pain, which is "life's norm."[17]

Verhaeren sets the stage in "Coincidence": "Oh! These beaches of the North, as I feel their winds from the West sweep over me the memory of hurricanes and clouds."[18] He dedicates the poem "The Rocks" to the people of Brittany, who are "clothed by the tide and battered by the storm." They are forced "to dream of some architecture in honor of the place – and of the wind; decisive strength in towers and crenels of solitude."[19] In his collection, Verhaeren does not neglect "The Windmill," which is so concerned with the wind. It turns again and again and then dies:

> The windmill turns in the dead of night, very slowly,
> Under a sad and melancholic sky,
> It turns and turns, and its arms, the color of sediment,
> Is sad and weak and noisy, and tires infinitely.[20]

Besides, there are "crosses at the crossroads . . . where cries of space and flight hang in the air, cries and tatters of the wind in the great forest." When the storm blows, the mighty bridges are "fastened by iron but worked over by the wind."[21] In "The Towns," Verhaeren invokes "my foolish, windy soul," meaning in particular the black wind.[22]

The poem entitled "The Wind" deserves to be cited in full, but we will quote just two stanzas:

> The wind with his trumpet that heralds November;
> Endless and infinite, crossing the downs,
> Here comes the wind
> That teareth himself and doth fiercely dismember;
> With heavy breaths turbulent smiting the towns
> The savage wind comes, the fierce wind of November.
> . . .
> The wind, it sends scudding leaves from the birches
> Along o'er the water, the wind of November,
> The savage, fierce wind;
> The boughs of trees for the birds' nests it searches
> To bite them and grind.
> The wind, as though rasping down iron, grates past.[23]

In this collection alone, the citations concerning the wind are innumerable. They highlight, writes Christian

Berg, the powers – above all, the powers of the winds – that never cease to bite, to slash, to tear, to cut, to hollow out, to grind, to file down, and to pierce.[24] The wind, together with the destructive forces in nature, reveals to the poet the essence of pain, which it symbolizes.

10

Short Strolls in the Wind of the Twentieth Century

Let us go from one era to the next for a brief excursion into the twentieth century. This will be in the form of a sketch in order not to interrupt too harshly this stroll in the wind, or, if we prefer, the variations that are consecrated to it. In the literature from the last century, two names stand out when it is a question about the winds, whether because this word is in the title or is a part of the text: Saint-John Perse (1889–1975) and Claude Simon (1913–2005).

In the youthful work of the first writer, the wind is a vehicle for a poetry that celebrates an island setting. Here everything is clear, but what is hidden behind the title *Vents* [Winds]? That is what will concern us. It is necessary to be a bit timid and reserved about this, so much so that it seems difficult for commentators to respond to this question, and so we must show its great simplicity. Saint-John Perse wrote this text in 1945 on a small island off the coast of Maine, and it was published in 1949; it was therefore published rather late, after he returned to literature following his well-known career as a diplomat.

I agree with Paul Claudel (1868–1955), one of the many commentators on the author's work, that the relationship between the wind and the expanse of space is essential from the very outset. Saint-John Perse draws up a list of the great elemental forces covered by the winds, and he sets down the notions of the "great migratory winds," the "reservoirs of the incommensurable," and the "respiratory function of the world."[1] More specifically, with this spatial perspective in mind, he invokes the essentially American western wind, which he claims to be "the pure ferments of a prenatal shadow mature in the West."[2] According to Claudel, his poetry is "nourished by space"; it is a devourer of space; however, he adds: "The wind – and there is only one – comes from the west, with its constant variations in duration and intensity and its infinitely simple modulations in intentions."[3]

Also important in Saint-John Perse's text is the relationship between man, wind, and time. The allusions to youth and old age are clear, and so is the certainty, on its own terms, of the reversibility of the times, as well as the desire to restore the men of the wind from former times.

In reading Saint-John Perse's *Winds*, one lesson is no less clear in my mind: "The winds are strong! The flesh is brief! . . . "[4] So, therefore, in Saint-John Perse's opinion, ". . . we should live! With the torch in the wind, with the flame in the wind . . . "[5] And, furthermore, ". . . should the face of any man near us fail to do honor to life, let that face be held by force into the wind."[6]

Let us highlight once again this hymn to movement. The wind is inhabited by "the hub of forces yet untried," and it saves us from "a reflux in the desert of the instant."[7] The author's exaltation is to listen to the wind's lessons. He regrets the absence of the great book penetrated by the wind's thought. To end this lesson, let us listen to Saint-John Perse, who tells us: "Open your porches to the new Year! . . . A world to be born under your footsteps! . . . Hasten! Hasten! Word of the greatest Wind! . . . I shall hasten the rising of the sap in your acts. I shall lead your works to maturation."[8]

If we limit ourselves to the extent that the author's experience of the wind is clearly visible throughout this novel, to end these chapters devoted to the wind's imagery with Simon's book would be misleading. However, in my opinion, in so far as it shows the depth of our subject (i.e., the wind in a work of fiction), Simon's novel *Le Vent: tentative de restitution d'un retable baroque* [The wind: an attempt at the restoration of a baroque altar-piece] should find a place in the conclusion of this chapter.[9]

In this novel, the wind is the counterpoint of the plot, or, if we prefer, "the continual basis of the drama." The wind in question is the tramontane, which blows northwest across Languedoc and Roussillon. In this region, it drives a useless power, an anger without sense, object, or pretext, as Gaston Bachelard writes.[10] On the one hand, the wind manifests as a force that wreaks havoc on nature, notably on the vegetation that tries to resist it. It attacks the world's matter, from dust to sand. Its itinerary is simple: it is everywhere: in town, in the amphitheater, all are penetrated by its hard-hitting blows. It causes all objects to flee before it. It is an omnipresent noise, a tactile presence. It even prevents the lighting of a cigarette.

When it ceases, sometimes for a long time, when its groans and its tantrums stop, it persists in memory. It causes reminiscences. From this fact springs its characterization as the absence of temporal limits. According to Jean-Yves Laurichesse, the wind in Simon's novel is the great character of the times. It is the personification of the soul in pain, "condemned to waste away without end," without nourishing any hope of an end.[11] In this, Simon's black wind is the expression of the abyss of eternity. It is the polar storm in which the wandering Jew of Eugène Sue (1804–1875), condemned to follow the wind ceaselessly, appears on the borders of the earth.[12] And it is the gale that blows in the sails of the Ghost Ship that accompanies the same type of outcast.

Nevertheless, there is something more: in its assaults, its tantrums, its rows, and, above all, in its complaints, we could read the wind as a victim of an eternal damnation, condemned not to die, a reproach to humankind, who must accede to death.

11

The Wind, the Theater, and the Cinema

After the immense success of *Eidophusikon*, by Philip James de Loutherbourg (1740–1812), and other dioramas, theater directors had to attempt to satisfy a public becoming more and more difficult to please as far as the soundscape is concerned. This was especially true about the sound effects of the wind, because, in a number of plays, the spectator needed to hear and to feel the wind. What would a presentation of Shakespeare's *Macbeth* be without any wind on the witches' moor or when the king was murdered? A few drama critics in the nineteenth century were very observant, even difficult to please, as far as the wind's theatrical representation was concerned. This was certainly the case with Henrik Ibsen (1828–1906). Thus, the final scene in the play *Når vi døde vågner* [When we dead awaken] opens with the sound of the wind, and then the actor cries: "Do you hear these gales?" Several characters in the course of this play mention the weather: a strong gale, a strident gust of wind, a storm's growing intensity, etc.[1]

Figure 4 Wind machine. M. J. Moynet, *L'envers du théâtre: machines et decorations / The Other Side of the Theater: Machines and Decorations*, 1873.

All theaters in the nineteenth century had to make sure, when the need arose, that the blowing of the wind was heard and felt. Each one therefore possessed a machine more or less perfected for this purpose. In London theaters, especially those in Drury Lane, directors were particularly concerned with the soundscape. Even the more modest halls possessed a wind machine (figure 4). This machine was made up of "a cylinder mounted on a frame and covered with a canvas cloth." When it was time "to turn on the cylinder, the beating of the wooden slats against the canvas produced the sound of the wind. In varying the speed of the cylinder's rotation or the tautness of the cloth, the machine operator could play with the volume or the quality of the wind's sound."[2]

At the end of the century, the use of silk instead of canvas went a long way in allowing for the imitation of "the whistling of the wind that rushed through chimneys and corridors."[3] Most of these machines were of a small size, almost always portable. They were operated by qualified specialists who had to adapt to the scene unfolding on the stage, varying the speed and the rhythm of the cylinder's rotation and modifying the tautness of the slats or the cloth that imitated the wind's sound.

That said, the importance of the wind in the stagecraft of the theaters in the nineteenth century was weak compared to what would be conferred upon it by the cinema in the twentieth century. This medium had the power to welcome the wind and its movement and to cause the spectator to experience the enigmatic passage of the wind. As Benjamin Thomas has written: "The wind, when it is called upon in the cinematic image, seems to express its substance, to the point that we might say that *cinema is the wind*. Both share an essential motion that moves the objects that it surrounds, touches, or crosses." In other words, again citing Thomas, the cinema aims for "the fascinating beauty of the elements' free variations." That is why it is appropriate to talk about the cinema's wind-loving nature.[4]

Since the birth of cinema – and by this we mean the motion picture *Repas de bébé* [Baby's dinner] by the Lumière brothers in 1895 – the wind has become "the first breath of the world in images."[5] When it appears in a film, the wind introduces a moment of reality, an indication that contrasts with the fiction that is presented in the story line. It introduces into cinema a truth that is both inhuman and enigmatic, as we already know, and, paradoxically, it is a truth indifferent to human events. The wind, as Élisabeth Cardonne-Arlyck claims, is the first movement in which the real is affirmed.[6] According to Thomas, "the wind explains the power that film has in constructing a world."[7] All of this takes place even before the plot can unfold.

Of course, the wind's forms and functions differ according to the film. Sometimes, in making itself a monster of meteorological catastrophe, the gale, or the wind's extreme force, brings death and madness. It is the transference into cinema of the "bad wind" from literature that Hugo illustrated so well in "Gastibelza," a poem sung by a man driven crazy by the wind that blows across the mountains.[8]

It so happens that the wind in cinema is allegorical. It symbolizes a force coming from the outside: maybe from the north, or from the "Beyond," or from a revolution. In the last case, it is made into a symbol of a new era. With the same perspective in mind, the wind can impose itself as the master of the place. This is the case in *The Thin Red Line*, directed by Terrence Malick (b. 1943), which takes place on Guadalcanal in World War II. Another eventual role of the wind in cinema is when gusts become monsters that symbolize an event or cause a crisis or dress themselves up with a "apocalyptic aura."[9]

There remains one essential question: How does one film the wind? This is what Joris Ivens (1898–1989) demanded of himself when he directed *A Tale of the Wind* in 1988 (figure 5). The soundtrack could reinforce the images of the wind, which blows, breaks, or carries away objects; and then the director could focus on the

Figure 5 Joris Ivens and Marceline Loridan, *Une histoire de vent / A History of the Wind*, 1988.

© Bridgeman Images.

reactions of men and women under the wind's assaults, as Jean Epstein (1897–1953) did in *Le Tempestaire* [The storm tamer]. Or the director could linger on the wind's pure movement in a thousand different forms, as Thomas again notes: a squall that carries away "the dust at the top of a hill," a "horse with a dancing mane," eyes that screw up in the face of the wind, trees that sway in the wind, or ripples on the surface of water.[10]

In Epstein's film, succeeding shots show "a man confronting the wind while holding his hat in his hands; two gendarmes whose bearing is challenged by the mistral; a bride whose train is rendered useless by a squall; and a young woman who has to tamp down her dress because the wind is constantly trying to lift it up."[11] Each shot reminds us of the scene in which artificially blown air – because there is no question of a breeze in this photograph – lifts up Marilyn Monroe's dress.

There are subtler things, too: the encounter, in one way or another, of beauty caressed or accentuated when it is confronted by the wind's constant presence in surrounding foliage. There is also Monica Vitti's radiant beauty in *L'Eclisse* [The eclipse], directed by Michelangelo Antonioni (1912–2007).[12]

One scene that is recurrent throughout the history of the wind in cinema is the drying of clothes on a clothesline. In such scenes, it might be said that the wind dresses itself in clothing. By its presence alone, it excites in the spectator a range of feelings and reminiscences which associate the wind with clothing, cleanliness, domestic work, and sexuality. The scene is produced by different kinds of wind, from breezes to gales, and it is a symbol of either success or failure.

To conclude, let us emphasize the most beautiful gift of cinema for those who are interested in the history of the emotions aroused by the wind: more immediate than literature, music, or the fine arts, film is a medium that, in a variety of ways, is best suited to revealing – or, rather, producing – the feeling of "the mysterious passage of the winds."[13]

Postlude

Are there new experiences of the wind these days that could be added to those that we have revealed here? Are there a number of solitary explorers, launched upon journeys in many regions of the earth, who are in search of new sensations and emotions? The answer to this question can only be positive. And this is simply because our sensibility is no longer that of a previous age.

Here is the outline of another book that could be dedicated to the completely contemporary relationship between man and wind. It could be introduced by Jean-Paul Kauffmann's wonderful work entitled *L'Arche de Kerguelen: voyage aux îles de la Désolation* [The Kerguelen archipelago: a voyage to the Isles of Desolation], which was published in 1993. Kauffmann's travelogue could be considered as a sequel, just a century later, to the lived experiences of John Muir, but in an entirely different terrain. Similar to Muir's work, this book proclaims, page after page, that the wind in this collection of southern islands constitutes the main character. Let us listen to Kauffmann: "On the Kerguelen islands blows an unknown wind from other lands visited by this great traveler. In this valley that I thought was dead, it was revealed to me why the wind was at the origin of the creation of the world."[1]

The author says that he knows why the wind from the Kerguelen islands is unique: it never whistles. "No obstacle stands against it: neither trees, nor houses, nor electric wires, nor walls. It growls instead of sounding shrill notes to which we are accustomed in civilized lands. Its voice has the power of the chants of the Orthodox liturgy." Added to this is a sensation of an avalanche that descends from the heights. "I have the sensation," writes Kauffmann again, "that a moraine of gusts hurls against our backs. The earth shakes and I'm afraid."[2]

The author sketches out an unexpected relationship between wind and political power: "The wind governs the archipelago, even though officially the French are masters of the district. In the face of the wind, we dominate nothing. We can subjugate the burning desert, the icy expanses, or even the humid zones of the earth, but we cannot subjugate the wind." Like an echo from Ecclesiastes 8:8, "It is not in man's power to restrain the wind," Kauffmann writes: "In the Kerguelen islands, the wind proclaims the absolute fluidity of things. Time has no weight, and the future has no prospects." Here reigns "the absence of the viscosity of time."[3]

The entire book is a description of the wind's brutality. As for water, "annihilated by the squall, it disperses in droplets that flutter like a thousand fireflies." The wind's sovereignty adjusts itself to the higher altitudes. "In the silence of the heights, the wind blows with a strangled and painful voice. Its breathing is short and labored. It repeats and repeats in stifled contractions." On the mountain, "the wind is an organist whose worn-out body does the scales on basalt pipes with a regal fluidity." Here rules "the regular breathing of the universal bellows." In the Kerguelen islands, the attempt to create a whaling station failed: "The desiccating force" of the "Aeolian fire" dried everything up, and there no longer remains anything except a ruin.[4]

Page after page, Kauffmann has drawn up a list of the wind's forms, which, for my part, I have found in no other book on any other territory. In the Kerguelen islands, "the wind . . . growls with a voice coming from another world."[5]

Thus ends our promenade in the wind, the elementary force that has forever been at the heart of human experience, totally mysterious throughout the millennia, but bit by bit tamed, without erasing its dreamlike force. We get the feeling of its link with the world's origin and the breath of Creation. We recognize the way the wind acts as a messenger of oblivion and, in its intense profundity, as a premonition of death.

Notes

Prelude

1 Victor Hugo, *Oeuvres complètes*, ed. Paul Meurice (Paris: Ollendorff, 1904–24), vol. 3: *Les Miserables*, *Préface philosophique*, p. 324.

2 Joseph Joubert, *Recueil des pensées de M. Joubert*, ed. François-René de Chateaubriand (Paris: Le Normant, 1838), p. 323.

Chapter 1 The Inscrutable Wind

1 Horace Bénédict de Saussure, *Voyages dans les Alpes* (Geneva: Georg, 2002), pp. 237–8.

2 Ibid.

3 François Rabelais, *Gargantua and Pantagruel*, trans. C. M. Cohen (London: Penguin Books, 1955), bk 4, ch. 43, pp. 540–2.

4 This need for ventilation and the politics surrounding it is elaborated in Alain Corbin, *The Foul and the Fragrant: Odor and the French Social Imagination*, trans. Miriam L. Kochan (Cambridge, MA: Harvard University Press, [1982] 1986).

5 Alexander von Humboldt, *Cosmos: A Sketch of a Physical Description of the Universe*, trans. E. C. Otté (London: Bell & Daldy, 1858), vol. 1, p. 316.

6 Ibid., vol. 1, p. 347.

7 On all of these points, see Fabien Locher, *Le Savant et la tempête: étudier l'atmosphère et prévoir le temps au XIXe siècle* (Rennes: Presses Universitaires de Rennes, 2008); and Numa Broc, *Une histoire de la géographie physique en France (XIXe–XXe siècles)* (Perpignan: Presses Universitaires de Perpignan, 2010), vol. 1, pp. 187–202.

8 On Léon Brault and his work, see Locher, *Le Savant et la tempête*, pp. 154–9.

9 Humboldt, *Cosmos*, vol. 1, p. 306.

10 Jean-François Minster, *La Machine océan* (Paris: Flammarion, 1997), p. 48.

Chapter 2 The Winds of the Common Folk

1 Jules Michelet, *Mountain*, trans. W. H. Davenport Adams (New York: T. Nelson & Sons, 1872).

2 Jean-Pierre Destand, "Éole(s) en Languedoc: une ethologie sensible," *Ethnologie française* 39 (2009): 598–608.

3 Martine Tabeaud, Constance Bourtoire, and Nicolas Schoenenwald, "Par mots et par vent," in Alain Corbin (ed.), *La Pluie, le soleil, et le vent: une histoire de la sensibilité au temps qu'il fait* (Paris: Aubier, 2013), pp. 69–88.

4 Patrick Boman, *Dictionnaire de la pluie* (Paris: Seuil, 2007), p. 361.

5 Jean-Pierre Richard (1922–2019) was a literary critic, whose book *Littérature et sensations* (Paris: Seuil, 1954) explored the relationship between writers

(Stendhal, Flaubert, Fromentin, and the Goncourt brothers) and their experiences of the physical world. It is unclear whether Corbin is referring to this work or another work when he mentions Richard's analysis of weathervanes. Nor does he provide any references to the histories of windmills.

6 Alphonse Daudet, *Letters from my Mill*, trans. Katherine Prescott Wormsley (Boston: Little, Brown, 1901), "Master Cornille's Secret," pp. 10–11.

7 Ibid., "The Lighthouse," p. 55.

Chapter 3 The Aeolian Harp

1 Anouchka Vasak, "Héloïse et Werther, *Sturm und Drang*: comment la tempête, en entrant dans nos coeurs, nous a donné le monde," *Ethnologie française* 39 (2009): 677–85.

2 Samuel Taylor Coleridge, "The Eolian Harp," in *The Complete Poetical Works of Samuel Taylor Coleridge*, ed. Ernest Hartley Coleridge (Oxford: Clarendon Press, 1912), vol. 1, pp. 100–2.

3 Pauline Nadrigny, "L'Écho des bois: une création originale de la Nature," in Jean Motter (ed.), *La Forêt sonore: de l'esthétique à l'écologie* (Seyssel: Champvallon, 2017), p. 60.

4 Johann Wolfgang von Goethe, "Äolsharfen," in *Gesammelte Werke in Sieben Bänden*, ed. Bernt von Heiseler (Weimar: Bertelsmann, 1961).

5 Maine de Biran, *Journal*, ed. Henri Gouhier (Neuchâtel: La Baconnière, 1957), vol. 3, p. 33.

6 Henry David Thoreau, *Journal*, ed. Robert Sattelmeyer et al. (Princeton, NJ: Princeton University Press, 1981–2002), vol. 3, p. 323 (July 21, 1851), vol. 4, p. 143 (October 12, 1851).

7 Eugène Delacroix, *Journal, 1822–1863* (Paris: Plon, 1980), p. 751.

8 Cited in Marine Ricord, "'Parler de la pluie et du beau temps' dans la *Correspondance* de Mme de Sévigné," in Karin Becker (ed.), *La Pluie et le beau temps dans la littérature française: discours scientifiques et transformations littéraire du Moyen Âge à l'époque moderne* (Paris: Éditions Hermann, 2012), pp. 174, 175, 179.

9 Jean-Jacques Rousseau, *La Nouvelle Héloïse: Julie, or, The New Eloise*, trans. Judith H. McDowell (University Park: Pennsylvania State University Press, 1968).

10 Anouchka Vasak, "Naissance du sujet moderne dans les intempéries: météorologie, science de l'homme et littérature au crépuscule des Lumières," in Becker, *La Pluie et le beau temps*, pp. 237–55.

11 Ibid., p. 251.

12 Alain Corbin, *The Lure of the Sea: The Discovery of the Seaside in the Western World, 1750–1840*, trans. Jocelyn Phelps (Berkeley: University of California Press, 1994).

13 Ibid., p. 41.

14 Anthony Reilhan, *A Short History of Brighthelmston* (London: W. Johnstone, 1761; London: Philanthropic Society, 1829), pp. 24–5.

15 Richard Townley, *A Journal Kept in the Isle of Man* (Whitehaven: J. Ware & Son, 1791). See especially (vol. 1) May 12, 1789, p. 17; July 19, 1789, p. 116; August 4, 1789, pp. 137–8; and April 5, 1789, p. 9. See also the entries of May 13, 1789, p. 17, and August 27, 1789, p. 166.

16 Bernardin de Saint-Pierre, *Studies of Nature*, trans.

Henry Hunter (Philadelphia: Joseph J. Woodward, 1836), pp. 307, 395–6. The final phrase in the last quotation was not included in Hunter's translation.

17 François-René de Chateaubriand, *Memoirs of Chateaubriand: From his Birth in 1768 till his Return to France in 1800* (London: Henry Colburn, 1849), pp.67, 128, 137, 136, 138, and 139.

18 François-René de Chateaubriand, *Atala and René*, trans. A. S. Kline (Scotts Valley, CA: CreateSpace, 2010, 2011), pp. 11 and 20 [online].

19 Alain Corbin, "Les émotions individuelles et le temps qu'il fait," in Alain Corbin, Jean-Jacques Courtine, and Georges Vigarello (eds), *Histoire des émotions* (Paris: Seuil, 2017), pp. 43–57, especially p. 49.

20 Maine de Biran, *Journal*, vol. 3, pp. 48, 49, 52.

21 Ibid., vol. 3, pp. 83, 85, 86.

22 Claude Reichler, "Météores et perception de soi: un paradigme de la variation liée," in Becker, *La Pluie et le beau temps*, pp. 213–36.

23 Maurice de Guérin, *Oeuvres completes*, ed. Marie-Catherine Huet-Brichard (Paris: Classiques Garnier, 2012), p. 62.

24 Ibid., pp. 85–6; Maurice de Guérin, *Journal of Maurice de Guérin*, ed. G. S. Trébutien, trans. Edward Thornton Fisher (New York: Leypoldt & Holt, 1867), pp. 105–6 (December 21, 1833).

25 Ibid., pp. 79–80, 100–1 (December 8, 1833).

26 Ibid., pp. 63–4, 82–3 (May 1, 1833).

Chapter 4 New Experiences of the Wind

1 Anouchka Vasak, *Météorologies: discours sur le ciel et le climat des Lumières au romantisme* (Paris:

Champion, 2007), ch.1, "L'Orage du 13 juillet 1788," pp. 37ff.

2 Cited ibid., p. 78.

3 Ibid., p. 85.

4 Ibid., p. 87.

5 Ibid., p. 95.

6 Louis Antoine de Bougainville, *A Voyage around the World*, trans. John Reinhold Forster (London: J. Nourse, 1772), pp. 131–2.

7 Ibid., p.188.

8 Ibid.

9 Anouchka Vasak, "Joies du plein air," in Guilhem Farrugia and Michel Delon (eds), *Le Bonheur au XVIIIe siècle* (Paris: Presses Universitaires de Rennes, 2015), pp. 193ff. The most comprehensive work on the history of ballooning is Marie Thébaud-Sorger, *L'Aérostation au temps du Lumières* (Rennes: Presses Universitaires de Rennes, 2009), especially pp. 247–53. See also her article "La conquête de l'air, les dimensions d'une découverte," *Dix-huitième siècle* 31 (1999): 159–77.

10 Raphaël Troubac, *Le Théatre que des hommes voyaient pour la première fois*, thesis, Université Paris I Panthéon-Sorbonne, 1999. This academic study is the closest to our project, and thus we largely rely on it concerning impressions and sensations.

11 Ibid., pp. 8ff.

12 Ibid., pp. 35–6.

13 See Fabien Locher, *Le Savant et la tempête: étudier l'atmosphere et prévooir le temps au XIXe siècle* (Rennes: Presses Universitaires de Rennes, 2008), pp. 169ff.

14 Guy de Maupassant, *En l'air et autres chroniques*

d'altitude (Paris: Éditions du Sonneur, 2019), pp. 37, 41, and 42.

15 Ibid., p. 45.

16 Ibid., p. 49.

17 René Caillié, *Travels through Central Africa to Timbuctoo; and across the Great Desert to Morocco* (London: Henry Colburn & Richard Bentley, 1830), vol. 2, p. 114.

18 Ibid., p. 115.

19 Guy Barthélemy, *Fromentin et l'écriture du désert* (Paris: L'Harmattan, 1997), p. 27.

20 Cited ibid.

21 Pierre-Marc de Biasi, "Présentation," in Gustave Flaubert, *Voyage en Égypte* (Paris: Grasset, 1991).

22 Gustave Flaubert, *Flaubert in Egypt: A Sensibility on Tour: A Narrative drawn from Gustave Flaubert's Travel Notes and Letters*, trans. Francis Steegmuller (Boston: Little, Brown, 1972), pp. 181–2.

23 Ibid., pp. 182–3.

24 Jules Verne, *Five Weeks in a Balloon; or, Journeys and Discoveries in Africa by Three Englishmen*, trans. "William Lackland" (New York: Hurst, 1869), pp. 217–18.

25 Ibid., p. 218.

26 Henry David Thoreau, *Cape Cod* (Boston: Houghton Mifflin, 1914), pp. 245–6.

27 Barbara Maria Stafford, *Voyage into Substance: Art, Science, Nature and the Illustrated Travel Account, 1760–1840* (Cambridge, MA: MIT Press, 1984).

28 John Muir, *Mountains of California* (New York: Century, 1894), "A Wind-Storm in the Forest," p. 255. Corbin relies on an anthology of Muir's

writings translated into French; see John Muir, *Célébration de la nature* (Paris: José Corti, 2011).

29 John Muir, *Our National Parks* (Boston: Houghton, Mifflin, 1901), "The Sequoia and General Grant National Park," p. 283.

30 John Muir, *Steep Trails: California, Utah, Nevada, Washington, Oregon, The Grand Canyon* (Boston: Houghton, Mifflin, 1918), "A Great Storm in Utah," pp. 116–17.

31 Ibid., pp. 114–15.

32 Muir, *Mountains of California*, p. 244.

33 Ibid., pp. 247, 249.

34 Ibid., pp. 249–50.

35 Ibid., p. 250.

36 Ibid.

37 Ibid., p. 254.

38 Ibid.

39 Ibid., p. 253.

40 Ibid., p. 254.

41 John Muir, *The Story of My Boyhood and Youth* (Boston: Houghton Mifflin, 1913).

42 Muir, *Mountains of California*, p. 255.

Chapter 5 The Tenacity of the Aeolian Imagination in the Bible

1 Corbin uses the French version of the *Bible of Jerusalem*; the translator is using the *New English Bible*.

2 Genesis 1:2.

3 Genesis 8:1.

4 1 Kings 19:11.

5 1 Kings 19:12.

6 Job 27:12.

7 Job 1:4.

8 Job 55:8.

9 Psalms 18:10.

10 Psalms 18:15.

11 Psalms 50:3. The *New English Bible* translates the subject of this verse as "a consuming fire"; The *Jerusalem Bible* translates it as "a raging wind."

12 Psalms 77:26.

13 Psalms 104:3–4.

14 Psalms 135:5, 7.

15 Psalms 148:8.

16 Psalms 107: 23, 25–9.

17 Ecclesiastes 1:6.

18 Ecclesiastes 1:14.

19 Wisdom of Solomon 4:3–4.

20 Wisdom of Solomon 5:14.

21 Wisdom of Solomon 5:23.

22 Ecclesiasticus 39:28.

23 Ecclesiasticus 43:17.

24 Jeremiah 4:11–12.

25 Jeremiah 10:13.

26 Jeremiah 49:36.

27 Ezekiel 13:13.

28 Prayer of Azariah 1:26. Catholics include this poem in chapter 3 of Daniel. Protestants relegate it to the Apocrypha.

29 Prayer of Azariah 1:43.

30 Nahum 1:2.

31 Zechariah 6:5–7.

32 Zechariah 6:8.

33 Matthew 8:26–7.

34 Mark 4:37, 39.

35 Matthew 14:30, 33.

36 Mark 13:27.

37 John 6:16–21.

38 Acts 2:3 and 2:2.

39 Acts 27:14.

40 Revelation 7:1.

Chapter 6 The Epic Power of the Wind

1 Homer, *The Odyssey of Homer*, trans. Richmond Lattimore (New York: Harper & Row, 1975), Book X, lines 19–26, p. 152.

2 Ibid., Book X, lines 46–9, p. 153.

3 Ibid., Book V, lines 295–6, 330–2, pp. 95–6.

4 Ibid., Book V, lines 383–5, p. 98.

5 Ibid., Book IV, lines 564–8, pp. 79–80.

6 See Violaine Giacomotto-Charra, "Le 'magazine des vents': les enjeux de l'exposé météorologique dans *La Sepmaine* de Du Bartas," in Karin Becker (ed.), *La Pluie et le beau temps dans la littérature française: discours scientifiques et transformations littéraire du Moyen Âge à l'époque moderne* (Paris: Éditions Hermann, 2012), pp. 147, 149ff.

7 Ibid., p. 158.

8 John Milton, *Paradise Lost*, ed. Gordon Teskey (New York: W. W. Norton, 2005), Book V, lines 15–18, 192–3, 654–5, pp. 106, 111, 124.

9 Ibid., Book IX, lines 199–200, p. 202.

10 Ibid., Book X, lines 695–707, 710–11, p. 249.

11 Ibid., Book X, lines 712–13, p. 249.

12 Ibid., Book X, line 694, p. 249.

13 Corbin uses the following French translation: Friedrich Gottlieb Klopstock, *La Messiade* (Paris: Hachette, 1849); the translator is using the following English version: *The Messiah: A Poem*, trans. F. A. Head (London: Longmans, 1826).

14 Bach's *St John Passion* was written in 1724 and *St Matthew Passion* in 1727. Corbin may have been thinking of Handel's *Messiah*, written in 1742.

15 Ibid., vol. 1, p. 72; lines 361–3, 373–4, 357–9, p. 59.

16 Jean-Baptiste Cousin de Grainville, *Le Dernier Homme* (Paris: Payot, 2010); based on the 1811 edition edited by Charles Nodier with a preface by Jules Michelet. See the English translation *The Last Man*, trans. I. F. and M. Clarke (Middletown, CT: Wesleyan University Press, 2002).

17 Ibid., p. 91.

18 Corbin uses the French translation Le Tasse, *La Jérusalem délivrée* (Paris: Classique Garnier, 1990); the translator is using *The Jerusalem of Torquato Tasso*, trans. J. H. Wiffen (New York: Appleton, 1858).

19 Ibid., Canto 3, stanza 76, p. 142.

20 Ibid., Canto 1, stanza 90, p. 91.

21 Ibid., Canto 18, stanza 85–6, p. 530.

22 Ibid., Canto 18, stanza 86, p. 530.

23 Ibid., Canto 9, stanza 52, p. 302.

24 Ibid., Canto 13, stanza 61, p. 407.

25 Ibid., Canto 14, stanza 19, p. 421.

26 Pierre de Ronsard, *The Franciad (1572)*, trans. Phillip John Usher (New York: AMS Press, 2010), lines 71–7, p. 76.

27 Ibid., lines 81, 87, p. 76.

28 Ibid., lines 89–94, pp. 76–7.

29 Ibid., lines 151–60, p. 78.

30 Ibid., lines 217, 219, 221, 231, 243, 253, pp. 80–1.

31 Corbin uses the French translation Luis de Camões, *Les Luisades ou Les Portugais*, trans. J. B. J. Millié (Paris: Firmin Didot, 1825); the translator is using

The Lusiads, trans. William C. Atkinson (London: Penguin Books, 1952), which, however, is an abridgement.

32 Ibid., Canto 1, p.44.

33 Ibid., Canto 1, p.48.

34 Ibid., Canto 5, p.128.

35 Ibid., Canto 5, p.125-126.

36 This phrase is in the French edition (see *Les Luisades*, vol. 1, p. 307), but not in the English translation: Canto 5, p. 134.

37 Ibid., Canto 6, pp. 147, 148.

38 Ibid., Canto 6, p. 157.

39 Ibid., Canto 6, p. 158.

40 Ibid., Canto 9, p. 203.

41 Luis Camões, *La Luisiade de Louis Camoëns: poème héroïque en dix chants*, trans. Jean-François de La Harpe (Paris: Nyon ainé, 1776).

Chapter 7 The Fantasy of the Wind in the Enlightenment

1 Yvon Le Scanff, *Le Paysage romantique et l'expérience du sublime* (Seyssel: Champ Vallon, 2007).

2 Cited ibid., p. 38.

3 François-René de Chateaubriand, *The Genius of Christianity*, trans. Charles I. White (Baltimore: John Murphy, 1856), p. 478.

4 Le Scanff, *Le Paysage romantique*, p. 27.

5 Ibid., pp. 31, 38.

6 James Macpherson, *Fragments of Ancient Poetry* (Edinburgh: G. Hamilton & J. Balfour, 1760), Fragment XI, p. 50.

7 Le Scanff, *Le Paysage romantique*, p. 43.

8 Macpherson, *Fragments*, Fragment III, pp. 16–17.

9 Ibid., Fragment VIII, p. 40.

10 Ibid., Fragment X, pp. 46, 48, 49.

11 Ibid., Fragment XII, pp. 55, 57.

12 Ibid., Fragment XIV, pp. 62, 64.

13 James Thomson, *Seasons* (London: A. Millar, 1767), "Winter," lines 111–17, p. 169.

14 Ibid., lines 195–7, pp. 171–2.

15 Ibid., lines 722–5, p. 190.

16 Ibid., lines 895–901, p. 197.

17 Ibid., "A Hymn," lines 1–3, 8–9, p. 204.

18 Ibid., lines 18–20, p. 205.

19 Ibid., lines 40–4, pp. 205–6.

20 Ibid., lines 45–7, p. 206.

Chapter 8 Gentle Breezes and Caressing Currents

1 Véronique Adam, "Écho aux quatre vents – la poétique de l'air baroque (1560–1640)," in Michel Viegnes (ed.), *Imaginaires du vent* (Paris: Imago, 2003), pp. 203–15.

2 Cited ibid., p. 204.

3 Ibid.

4 Régine Detambel, *Petit éloge de la peau* (Paris: Gallimard, 2007), pp. 121, 125, 128.

5 James Thomson, *Seasons* (London: A. Millar, 1767), "Spring," lines 258–9, p. 12.

6 Ibid., lines 322–4, p. 14.

7 Ibid., lines 475–6, p. 20.

8 Ibid., lines 498–9, p. 21.

9 Ibid., "Autumn," lines 515–18, p. 133.

10 Ibid., "Summer," lines 123, 5, 4, 1299, 1284, 1290–1, 1301–2, 1318–19, pp. 51, 47, 95, 94, 95.

11 Corbin uses the French translation: Salomon Gessner,

Contes moraux et nouvelles idylles (Zurich: Orell, Gessner, Fuessli, 1773); the translator is using *New Idylles*, trans. William Hooper (London: S. Hooper, Ludgate Hill, 1776).

12 Ibid., Idyl 7, pp. 16–17.

13 Ibid., Idyl 9, p. 25.

14 Leconte de Lisle, *Poèmes antiques* (Paris: Gallimard, 1994), p. 252.

15 Ibid., p. 253.

16 Ibid., p. 254.

17 Ibid., p. 373.

18 Edgard Pich, "Introduction," in Leconte de Lisle, *Oeuvres complètes*, ed. Edgard Pich (Paris: Belles Lettres, 1976), vol. 1, *Poèmes antiques*, pp. i–xlii.

19 Gustave Flaubert, *The Temptation of St. Anthony*, trans. Lafcadio Hearn (New York: Alice Harriman, 1910), p. 65.

20 Jean Giono, *Harvest*, trans. Henri Fluchère and Geoffrey Myers (San Francisco: North Point Press, 1983), pp. 64–5.

21 Ibid., pp. 65–7.

22 Ibid., pp. 99ff.

Chapter 9 The Enigma of the Wind in the Nineteenth Century

1 Victor Hugo, *The Works of Victor Hugo* (Boston: Jefferson Press, c.1900), vol. 9, *Poems, Dramas*, ed. Henry Llewellyn Williams, "Poor Folk," trans. Bishop Alexander, p. 250.

2 Yvon Le Scanff, *Le Paysage romantique et l'expérience du sublime* (Seyssel: Champ Vallon, 2007), pp. 88–9.

3 Victor Hugo, *The Toilers of the Sea*, trans. Isabel

F. Hapgood (New York: Signet Classic, 2000), pp. 359, 386, 12.

4 Victor Hugo, *The Man Who Laughs* (New York: J. H. Sears, c. 1900), pp. 77–8.

5 Hugo, *Toilers of the Sea*, p. 360.

6 Le Scanff, *Le Paysage romantique*, pp. 88–9.

7 Françoise Chenet, "Hugo ou l'art de déconcerter les anémomètres," in Michel Viegnes (ed.), *Imaginaires du vent* (Paris: Imago, 2003), pp. 297–309, especially p. 298.

8 Ibid., p. 304. Chenet cites Hugo's poem "La mer et le vent."

9 Ibid., p. 305.

10 Ibid., p. 307.

11 Victor Hugo, *Les Contemplations*, ed. Gabrielle Chamarat (Paris: Pocket, 1990), p. 499.

12 Leconte de Lisle, *Poèmes tragiques* (Paris: Lemerre, 1866), pp. 74–5.

13 Ibid., pp. 117–18.

14 Ibid., p. 125.

15 Ibid., p. 77.

16 Ibid., p. 78.

17 Werner Lambersy, "Préface," in Émile Verhaeren, *Les Villages illusoires* (Brussels: Communauté Française de Belgique, 2016), p. 5.

18 Verhaeren, *Les Villages illusoires*, p. 103.

19 Ibid., p. 120.

20 Ibid., p. 23.

21 Ibid., pp. 45, 51.

22 Ibid., p. 59.

23 Ibid., pp. 164–6. For the English translation, see *Poems of Emile Verhaeren*, trans. Alma Strettell (London: John Lane, 1915), pp. 43–4.

24 Christian Berg, "Postface," in Verhaeren, *Les Villages illusoires*, p. 215.

Chapter 10 Short Strolls in the Wind of the Twentieth Century

1 Paul Claudel, "Commentaire," in Saint-John Perse, *Oeuvres complètes* (Paris: Gallimard, 1982), pp. 1121–30, at p. 1130.
2 Saint-John Perce, *Winds*, trans. Hugh Chisholm (New York: Pantheon Books, 1961), Canto 1, p. 47.
3 Claudel, "Commentaire," p. 1122.
4 Saint-John Perce, *Winds*, Canto 2, p. 85.
5 Ibid., Canto 3, p. 121.
6 Ibid., Canto 1, p. 33.
7 Ibid., Canto 3, p. 111, and Canto 4, p. 135.
8 Ibid., Canto 4, pp. 175, 177.
9 Claude Simon, *The Wind*, trans. Richard Howard (New York: Braziller, 1959).
10 Gaston Bachelard, *The Poetics of Space*, trans. Maria Jolas (New York: Orion Press, 1964).
11 Jean-Yves Laurichesse, "Le vent noir de Claude Simon," in Michel Viegnes (ed.), *Imaginaires du vent* (Paris: Imago, 2003), pp. 113–25, especially p. 121.
12 See the preface by Francis Lacassin consecrated to the "furious storm" and the "boreal tempest" in Eugène Sue, *Juif errant* (Paris: Robert Laffont, 1983), p. 16.

Chapter 11 The Wind, the Theater, and the Cinema

1 Robert Dean, "Ibsen, le designer sonore du théâtre au XIXe siècle," in Jean-Marc Larrue and Marie-Madeleine Mervant-Roux (eds), *Le Son du théâtre* (Paris: CNRS, 2016), pp. 165–80.

2 Ibid., p. 167.
3 Jules Moynet, *L'Envers du théâtre: machines et décorations* (Paris: Hachette, 1875), p. 169.
4 Benjamin Thomas, *L'Attrait du vent* (Crisnée, Belgium: Yellow Now, 2016), pp. 13, 14, 18.
5 Élisabeth Cardonne-Arlyck, "Passages du vent au cinéma," in Michel Viegnes (ed.), *Imaginaires du vent* (Paris: Imago, 2003), pp. 126ff.
6 Ibid.
7 Thomas, *L'Attrait du vent*, p. 49.
8 Victor Hugo, "Gastibelza," trans. Nelson R. Tyerson, in Hugo, *The Works*, vol. 9, pp. 140–2.
9 Cardonne-Arlyck, "Passages du vent au cinéma," pp. 129, 134.
10 Thomas, *L'Attrait du vent*, pp. 74–5.
11 Ibid.
12 Ibid., pp. 56–9.
13 Cardonne-Arlyck, "Passages du vent au cinéma," p. 136.

Postlude

1 Jean-Paul Kauffmann, *L'Arche des Kerguelen: voyage aux îles de la Désolation* (Paris: Flammarion, 1993), p. 75.
2 Ibid., pp. 76, 90–1.
3 Ibid., pp. 115, 118.
4 Ibid., pp. 137–8.
5 Ibid., p. 169.

Index

Page numbers in *italics* refer to a figure in the text.

Index

Index

emotion 28–37, 43–6, 53–8
emptiness 43–4, 62
euphoria 44
fluidity 43, 47, 125
melancholy 27, 29, 90, 112
and memory 57, 89–90, 117
and natural theology 28, 29
peacefulness 36, 45, 47
rage 64, 80, 81, 91, 107, 108
tranquillity 4, 45, 94
emperors, French 13, 87
emptiness 43–4, 62
England 10, 29, 30
English people 28
enigmas 106, 120–1
Enlightenment 10, 29, 43, 75,
86–94
epics, Greco-Roman 59, 68–85
see also four winds
epidemics/disease 8, 10–11
Epstein, Jean 122–3
eroticism 96, 101–3
eternity 49, 117
euphoria 44
Eurus (east wind) 68, 70, 73, 78
experiencing the wind 38–58
see also ballooning; sandstorms;
sea, wind at; sequoias
Ezekiel, Book of 64

filming the wind 121–3
Flammarion, Camille 45–6
Flaubert, Gustave 50–1, 101
fluidity 43, 47, 125
force, wind as 7–8, 50, 56, 105,
116, 112, 126
four winds 63–4, 65–6, 68–70
Aquilon (north wind) 78, 79
Boreas (north wind) 68, 70, 73,
77, 84, 90
Eurus (east wind) 68, 70, 73, 78
Notus (south wind) 68, 70, 73,
77, 78
Camões 83, 84
Zephyrus (west wind) 68, 69, 70,
73, 77–8, 90
Camões 83, 85

Thomson 93, 94
see also east wind; north wind;
south wind; west wind
France 20, 22, 111–12
Franciad (Ronsard) 71, 78–80
Franck, César 100
Francus 78–9, 80
French navy 13–14, 16
French Revolution 39
fresh air, importance of 7, 8
Friedrich, Casper David 87

Galton, Francis 13
Genesis, Book of 60, 61
Gessner, Salomon 98–9
Giono, Jean 101–3
God
anger of 63–4, 65, 72, 73–4, 75,
77
in the Bible 62, 63–4, 65, 75
communication with 48, 49
faces of 93–4
instrument of 61, 63, 64, 65, 75,
77
punishment of 63, 73, 74
gods/goddesses
Aeolus 26, 68, 69, 79, 84,
99–100
Athena 68, 70
Neptune 79, 80, 83
Poseidon 68, 69–70
Venus 81, 84
Zeus 68, 69
Goethe, Johann Wolfgang von 26
Gothic novels 87
Grainville, Jean-Baptiste, Le Dernier
Homme 75–6
Grand Tour 28
Great Fear of 1789 39
Greco-Roman literature 59, 68–85
see also four winds
Greenwich Observatory 13
Guérin, Maurice de 33, 34–7
Gulf Stream 16–18

harbinger, wind as 39
harps, Aeolian 24–7, 57

Index

Index

Index

Index